Arbuscular Mycorrhizal Fungi

Arbuscular mycorrhizal fungi (AMF) are considered enormously important in contemporary agriculture and horticulture due to their important role in nutrient, biotic and abiotic stress management apart from enhancing plant health and soil fertility. AMF is one of the important fungi for soil aggregation, which helps in drought management. Hence, this book brings out an exclusive text on AMF for sustainable rice production. It provides comprehensive up-to-date knowledge on AMF in rice cultivation and for sustainable rice production in different ecologies without damaging the environment.

Salient Features:

1. Covers all the aspects of AMF in rice cultivation from diversity to applications
2. Documents AMF diversity based on metagenomic approach in rice ecosystems
3. Explains the importance of AMF in soil aggregation, which helps in drought management
4. Provides new unraveling knowledge about AMF for sustainable rice production in different ecologies without damaging the environment
5. Discusses the AMF role in induction of resistance in rice plants against some pests.

Arbuscular Mycorrhizal Fungi

For Nutrient, Abiotic and Biotic Stress Management in Rice

Edited by

Periyasamy Panneerselvam,
Pradeep K. Das Mohapatra,
Amaresh Kumar Nayak, Debasis Mitra,
Kulandaivelu Velmourougane and
Sergio de los Santos-Villalobos

CRC Press
Taylor & Francis Group
Boca Raton London New York

CRC Press is an imprint of the
Taylor & Francis Group, an **informa** business

Cover image: © Shutterstock

First edition published 2024
by CRC Press
6000 Broken Sound Parkway NW, Suite 300, Boca Raton, FL 33487-2742

and by CRC Press
4 Park Square, Milton Park, Abingdon, Oxon, OX14 4RN

Library of Congress Cataloging-in-Publication Data
Names: Panneerselvam, P., editor. | Mohapatra, Pradeep K. Das, editor. |
Nayak, A. K., 1969– editor. | Mitra, Debasis (Microbiologist), editor. |
Velmourougane, Kulandaivelu, editor. | Santos Villalobos, Sergio de los, editor.
Title: Arbuscular mycorrhizal fungi : for nutrient, abiotic and biotic stresses management in rice / edited by Periyasamy Panneerselvam, Pradeep Kumar Das Mohapatra, Amaresh Kumar Nayak, Debasis Mitra, Kulandaivelu Velmourougane and Sergio De Los Santos-Villalobos.
Description: First edition. | Boca Raton : CRC Press, 2023. |
Includes bibliographical references and index.
Identifiers: LCCN 2023002032 (print) | LCCN 2023002033 (ebook) |
ISBN 9781032406411 (hardback) | ISBN 9781032406435 (paperback) |
ISBN 9781003354086 (ebook)
Subjects: LCSH: Vesicular-arbuscular mycorrhizas. | Mycorrhizal fungi. |
Rice—Diseases and pests—Biological control. | Rice—Growth. | Plant-fungus relationships.
Classification: LCC QK918 .A73 2023 (print) | LCC QK918 (ebook) |
DDC 595.717/8—dc23/eng/20230417
LC record available at https://lccn.loc.gov/2023002032
LC ebook record available at https://lccn.loc.gov/2023002033

ISBN: 978-1-032-40641-1 (hbk)
ISBN: 978-1-032-40643-5 (pbk)
ISBN: 978-1-003-35408-6 (ebk)

DOI: 10.1201/9781003354086

Typeset in Times
by codeMantra

Contents

Preface

Arbuscular mycorrhizal fungi (AMF) are considered enormously important in contemporary agriculture and horticulture due to their important role in nutrient, biotic and abiotic stress management apart from enhancing plant health and soil fertility. Beneficial effects of AMF are being explored in most of the agricultural, horticultural, plantation and forestry crops, but not much in rice cultivation. There is a school of thought that AMF may not work under wetland conditions, but scientific evidence has proved that this fungal association is essential for rice cultivation to minimize the use of fertilizers and pesticides in addition to plant stress management. Rice is different from other agricultural crops since it is being cultivated in different ecosystems like shallow, deep, flooded water, aerobic and upland rainfed/irrigated conditions. Hence, the recommended packages or specific strains will vary from one to another ecosystem. Rice cultivation consumes a huge amount of nitrogenous and phosphatic fertilizers. To reduce the level of fertilizers, we must improve the N and P use efficiency. Many scientific findings proved that AMF interactions in rice significantly improve the nutrient use efficiency. Rice is also cultivated in coastal and polluted lands, wherein AM fungal applications are essential to mitigate salt and pollutant stresses. It is a well-known fact that rice cultivation needs a large amount of water to minimize this problem. Rice cultivation is being promoted in upland irrigated and aerobic conditions, where AMF association is essentially required for nutrient and drought management. AMF play an important role in induction of resistance in rice plants against some pests. AMF are one of the important fungi for soil aggregation, which helps in drought management. Recent findings indicated that AMF have some host preference in rice, and hence, selection and recommendation of the specific group of fungi are essential. Also, documentation of AMF diversity based on the metagenomic approach in the rice ecosystem is essential as it helps for selection of suitable fungi for specific ecology. Presently, there is no systematic recommendation of AMF for rice cultivation as it is being cultivated in different ecosystems. In view of the above, bringing out the consolidated or a whole package of AMF for rice cultivation is essential, and hence, this proposal is submitted to bring out an exclusive book for AMF for sustainable rice production. This book will provide comprehensive up-to-date knowledge on AMF in rice cultivation.

Editors

Dr. Periyasamy Panneerselvam graduated in agricultural science and earned his master's and doctoral degrees with gold medals in the field of Agricultural Microbiology from Tamil Nadu Agricultural University, Coimbatore. Starting his career at Central Coffee Research Institute, Chikmagalur, as a research assistant in 1998 and subsequently as a field scientist until 2006, he joined ICAR on 8 January 2007 as a scientist at the Indian Institute of Horticultural Research, Bengaluru. Then, he joined as a senior scientist at ICAR-National Rice Research Institute, Cuttack, India, on 24 September 2015. On September 2018, he was promoted as a principal scientist, and he continued his service in the same post. He was awarded as Fellow of the Confederation of Horticulture Associations of India in 2015 and FSASS in 2021. He received the Best NRRI Worker (Senior Scientist) award at ICAR-NRRI, Cuttack, in 2019 and Dr. B. Vasantharaj David Award in 2018 for his achievements in the field of agricultural microbiology. He has developed good agricultural practices (GAP) for on-farm processing of coffee for the minimization of mold/mycotoxin contamination and a novel microbial consortium and AM fungal package for nursery seedling production, root lesion nematode and leaf rust management. He has developed "Soilless Arbuscular Mycorrhizal Fungal Inoculum production," and a patent (3817/CHE/2014) has been granted for this invention. The Arka Microbial Consortium (AMC) developed by him is popular for horticultural crop production; its application saves 25% of NPK fertilizers, besides enhancing yields by 10%–16% in different horticultural crops. This AMC technology has been licensed to 45 entrepreneurs, including KVKs and the Department of Horticulture, Government of Karnataka, India. Another technology, i.e., Actinobacterial consortium (Arka Actino Plus), developed by him is playing a pivotal role in improving plant health in different horticultural and plantation crops, and this technology is also licensed to private entrepreneurs for the benefit of farming community. Recently, he has developed formulations such as "Microbial consortium (MC) for pest and disease management in rice-horticulture based cropping system in Sikkim," "Bio-fertilizer consortium (BC) for nutrient management in rice-horticulture based cropping system in Sikkim" and "NRRI decomposing microbial consortium for paddy straw residues management." As one of the co-investigators, he has developed the following technologies: Arka fermented Cocopeat, Azolla sporocarp formulation for nitrogen management in lowland rice, and liquid bioinoculant of endophytic (*Azotobacter chroococcum*) and rhizospheric (*Azotobacter vinelandii*) nitrogen-fixing bacteria for rice crop. He had received overseas training on bioremediation from the University of South Australia, under the NAIP program, in which he identified novel bacterial consortium for remediation of chromium contamination in soil and water. Dr. Panneerselvam has published 119 research and review papers in peer-reviewed journals and has authored 20 proceedings papers, 3 books, 40 popular articles and 8 research/technology bulletins. He has guided 15 postgraduate and 2 PhD students.

Dr. Pradeep K. Das Mohapatra obtained his PhD in 2008 from Vidyasagar University. Currently, he is working as an associate professor and Head in the Department of Microbiology, Raiganj University, Raiganj-733 134, Uttar Dinajpur, West Bengal, India. He is also acting as a Director of Professor A. K. Bothra Environment Conservation Centre, Raiganj University. Dr. Das Mohapatra has significantly contributed to the field of microbial enzyme biotechnology, environmental microbiology and industrial microbiology. He has worked on microbial nutrition and diversity, fermentation, probiotics, immobilization, bioinformatics and enzymes like tannase, xylanase, amylase, cellulase and keratinase. He is presently working on metabiotics. Dr. Das Mohapatra has published 3 books (*Microbial Fermentation and Enzyme Technology*, CRC Press, Taylor & Francis Group, and another 2 on biofertilizer), more than 150 original research papers in reputed *National and International Journals*, 8 book chapters, 5 review papers and 6 popular articles, and he has filed 2 patents. He has attended 8 training programs on microbial research, covered 40 national and international conferences, and delivered 22 invited lectures. Dr. Das Mohapatra has more than 20 years of research and 13 years of teaching experience, and he has received research funds from the UGC, Government of India, Biostadt India Ltd. and Sanzyme Private Limited. Under his supervision, 12 students were already awarded PhD, and 6 students are working currently to pursue their PhD. Dr. Mohapatra has received Jr. Scientists of the Year Award 2008 and Environmentalist of the Year Award 2009 from the National Environmental Science Academy, New Delhi. He was nominated as Fellow of the Society for Applied Biotechnology in 2011. He is serving as an editorial board member of three journals and a reviewer of many journals of Elsevier, Blackwell, Springer and Taylor & Francis. Dr. Das Mohapatra is also associated with many academic and research societies like AFB (France), NESA (New Delhi), BRSI (Trivandrum), AMI (New Delhi), PBVM (Kolkata), The Indian Science Congress Association (Kolkata) and SAB (Dharward).

Dr. Amaresh Kumar Nayak took charge as Director, ICAR-National Rice Research Institute, Cuttack, on 9 November 2022. He obtained his PhD in soil science and agricultural chemistry in 1997 from the Institute of Agricultural Sciences, BHU. He started his scientific career as a scientist at ICAR-CSSRI RRS, Anand, in 1996. Subsequently, he worked as a principal scientist at ICAR-PDFSR, Modipuram, and at ICAR-NRRI, Cuttack. From 2014 to till date, he has been working as the Head of the Crop Production Division at ICAR-NRRI. During his 25 years of research carrier, he has worked on problem soil management (saline, sodic and iron toxic soil), water quality mapping and modeling, management of poor quality irrigation water, carbon sequestration and nutrient management in rice-rice and rice-wheat systems, and the budgeting of greenhouse gas emissions in lowland rice-rice system under different water and nutrient management practices. He has been elected as Fellow of the National Academy of Sciences, India (FNASI); Fellow of the National Academy of Agricultural Sciences (FNAAS); Fellow of the Indian Society of Soil Science (FISS); and Fellow of the Association of Rice Researcher Worker (FARRW). He received many recognitions and awards like Nanaji Deshmukh ICAR Award 2020, SCON & SIT Award 2021, Endeavour Fellowship from the Department of Education and Training from the Government of Australia 2018, IPNI-FAI Award 2016, Hari Om Ashram Award 2012 from ICAR-New Delhi, ISSS-Dr. J.S.P. Yadav Memorial Award 2013 and Norman E. Borlaug Fellow in 2008. He was deputed as a visiting scientist to SWERI, ARC Egypt on salinity management in 2008 under the Indo-Egypt work plan. He has published over 219 research papers, 14 books, and several book chapters and technical bulletins for his credit.

Mr. Debasis Mitra received his B.Sc. degree in Biotechnology from Vidyasagar University, India, and his M.Sc. degree in Biotechnology from Graphic Era (Deemed to be University), India. He is an alumnus researcher of ICAR-Indian Institute of Horticultural Research, India. Currently, he is a doctoral scholar in the Department of Microbiology at Raiganj University, Raiganj-733 134, West Bengal, India, working at ICAR-National Rice Research Institute, India, and completed his virtual doctoral internship from Agroenvironmental and Health Biotechnology Laboratory, Universidad de Santander, Colombia. He earned online course certificates on "Microbiology and Forensic Science" from Jordan University of Science and Technology, Jordan;

"Climate Change" from Macquarie University, Australia; and "Agriculture and the World We Live In" from Massey University, New Zealand in 2018. He has published more than 70 research and review papers in various high-impact peer-reviewed journals of national and international repute, and he has contributed 3 books, 24 book chapters, 1 newsletter, 2 research technical bulletins, 1 extended summary, 3 technology and 100 e-Publication; is Editor/EBM of 6 journals; special edited 8 journals; and reviewed 130 papers (WoS) and several popular articles in leading magazines and journals. He is awarded as Fellow of the Scholars Academic and Scientific Society (FSASS) in 2021, received the best editor award of 2021 in *Indian Journal of Microbiology Research*, received the best reviewer award of 2022 in *Current Agriculture Research Journal* and *Biomedical and Pharmacology Journal* and ranked fifth (2022) and forth (2023) in World Scientist and University Rankings in 2022 and 2023 by AD Scientific Index. He has developed "Cocopeat with rice husk and saw dust compost as a nursery medium for growing vegetable seedlings." As one of the co-author, he has developed "Microbial consortium (MC) for pest and disease management in rice-horticulture based cropping system in Sikkim" and "Bio-fertilizer consortium (BC) for nutrient management in rice-horticulture based cropping system in Sikkim." He has attended more than 65 national and international conferences/webinar/seminar/workshop symposium/lectures. He is a member of the International Association for Agricultural Sustainability, American Association for Science and Technology, South American Mycorrhizal Research Network, GEO BON: Group on Earth Observations Biodiversity Observation Network, Society for the Study of Evolution and International Association of Oncology. He served as an editorial board member, special issue lead editor, guest editor and reviewer of several national and international journals. His research is focused on plant-microbe interactions, mycorrhizal interactions, soil microbiology, biocontrol, molecular biology and strigolactones.

Dr. Kulandaivelu Velmourougane obtained his PhD (gold medal) in microbiology from Indian Agricultural Research Institute (IARI), New Delhi. He is currently working as a senior scientist (microbiology) in the Division of Crop Production, ICAR-Central Institute for Cotton Research, Nagpur. Before joining ICAR as an ARS scientist, K. Velmourougane served as a scientist (microbiology) at the Central Coffee Research Institute, Coffee Board, Chikmagalur, Karnataka, working on the role of post-harvest technology in coffee quality improvement, including soil health, crop productivity and sustainable challenges in coffee production. He has wide experience in pesticide residues, pollution and mycotoxin management in coffee. Under the NAIP project "Georeferenced Soil Information System for Land Use Planning and Monitoring Soil and land Quality for Agriculture," he has worked on soil biological quality assessment of black soil regions and Indo-Gangetic plains of India, which covered 32 benchmark soils to study soil microbial diversity, soil enzyme activity and microbial biomass carbon, and linked these attributes to crop yield and soil quality index for the first time. He has worked on *Trichoderma viride–Azotobacter chroococcum* biofilm, its development and characterization. The developed biofilm has been proven beneficial in terms of plant growth enhancement, improving soil nutrient mobility and soil fertility in crops such as wheat, cotton and chickpea. Further, he has expertise in metagenomics and transcriptome analysis of biofilm and soil microorganisms. He underwent international training on "Carbon trading/carbon sequestration/climate change (crop sciences)" at the Center for Environmental Risk Assessment and Remediation (CERAR), University of South Australia, in 2014 under the guidance of Prof. Nanthi S. Bolan, Chair in Environmental Science, CERAR. Presently, he is working on the development of microbial biofilm-based formulations and rhizo-engineering for enhancing soil health and cotton productivity. He has been awarded a SERB-DST-funded project to work on microbial volatiles for sucking pest management in cotton. To his credit, he has more than 150 research publications, 3 copyrights, 3 training manuals, 14 technical reports/bulletins, 3 newsletters, 38 conferences and symposia, and 25 book chapters in national and international repute. His expertise in the subject is well evidenced from his publications in well-reputed international journals (*Journal of Basic Microbiology, Applied Microbiology and Biochemistry, Archives of Microbiology, Microbiological Research, Journal of Applied Microbiology, FEMS Microbial Ecology and Plant and Soil*). He has been a speaker in several national and international conferences to discuss various issues in coffee quality improvement, cotton microbiology, soil health enhancement, and role of microbial biofilms in soil and crop productivity.

Prof. Sergio de los Santos-Villalobos was born in Chahuites, Oaxaca, México, in 1985. He obtained his degree, with honorific mention, in Biochemistry Engineering from the Instituto Tecnológico de Celaya in 2007, where his scientific curiosity, in the area of biology, began, which was strengthened by his participation in the production of microbial inoculants of agricultural interest in the private sector. In 2012, he completed his doctorate at the Center for Research and Advanced Studies of the National Polytechnic Institute (CINVESTAV-IPN) Unidad Irapuato, with the specialty in plant biotechnology. In 2013 and 2014, he completed postdoctoral stays at CINVESTAV-Irapuato and the

Technological Institute of Sonora, respectively. He has conducted research stays in the United States, Chile, Brazil, Argentina, Paraguay, Cuba, New Zealand and Austria, among others, on cutting-edge topics in the agricultural biotechnology area. His current assignment, since 2014, is as Researcher of the National Council of Science and Technology commissioned to ITSON. He belongs to the National System of Researchers level 2. His scientific work has focused on the study of microbial ecology, plant x microorganism x soil interactions, as well as the production of microorganisms using agro-industrial residues. These lines of research have been developed with ecological, physiological, biochemical and genetic approaches (from in vitro to field trials). Since 2010, he has participated actively in international projects under the Regional Cooperation Agreement for the Promotion of Science and Nuclear Technologies in Latin America and the Caribbean, sponsored by the International Atomic Energy Agency, focused on the study of soil erosion and fertility. His scientific contributions have been honored with a special mention of the AgroBio 2013 (national) Prize, winner of the Tecnos 2014 (international) Award, obtaining patents and national registrations, member of the international expert panel of the International Atomic Energy Agency and Atheneum Partners on soil conservation and the generation of microbial inoculants for agriculture, respectively. He is currently the Head of the Microbial Resource Biotechnology Laboratory and Director of the Collection of Edaphic Microorganisms and Native Endophytes, Editor of the *Open Agriculture Journal*, *Current Research in Microbial Sciences* and *Chemical and Biological Technologies in Agriculture*. The lines of research developed and their products are the result of solid teamwork composed of students and collaborating institutions, to generate scientific knowledge and the integral training/strengthening of human resources.

Contributors

Nurudeen Olatunbosun Adeyemi
Department of Plant Physiology and Crop Production
Federal University of Agriculture
Abeokuta, Nigeria

Priyanka Adhikari
Center for Biodiversity Conservation and Management
G.B. Pant National Institute of Himalayan Environment
Uttarakhand, India

Wiem Alloun
Laboratory of Mycology, Biotechnology and Microbial Activity (LaMyBAM)
Department of Applied Biology
University Constantine 1
Constantine, Algeria

Suchismita Behera
Crop Production Division
ICAR-National Rice Research Institute
Cuttack, India

Ankita Behura
Department of Botany
Utkal University
Bhubaneswar, India

Shreosi Biswas
Division of Agricultural Chemicals
Indian Agricultural Research Institute
New Delhi, India

G. Capoani
Department of Life, Health and Environmental Sciences
University of Milan
Milano, Italy

Manju Chaithra
Division of Plant Pathology
ICAR-Indian Agricultural Research Institute
New Delhi, India

Debashree Dalai
Crop Protection Division
ICAR-National Rice Research Institute
Cuttack, India

Anamika Das
Department of Genetics and Plant Breeding
Bidhan Chandra Krishi Viswavidyalaya
West Bengal, India

Amar Jyoti Das
Environmental Microbiology Research Group
Department of Life Science
Graphic Era (Deemed to be University)
Uttarakhand, India

Thomas T. T. Dhanya
Department of Botany
Carmel College
Kerala, India

R. Djebaili
Department of Life, Health and Environmental Sciences
University of L'Aquila
Palazzo Camponeschi, Italy

Amit Kumar Dutta
Amity Institute of Biotechnology
Amity University
Jharkhand, India

Oni Olanrewaju Emmanuel
Department of Plant Physiology and Crop Production
Federal University of
Agriculture
Abeokuta, Nigeria

B. Farda
Department of Life, Health and Environmental Sciences
University of L'Aquila
Palazzo Camponeschi, Italy

Izdihar Ferhat
Laboratoire de Biotechnologies
Ecole Nationale Supérieure de Biotechnologie
Ville Universitaire Ali Mendjeli
Constantine, Algeria

Mahwish Firdous
Amity Institute of Biotechnology
Amity University
Jharkhand, India

Debashish Ghosh
Material Resource Efficiency Division
CSIR-Indian Institute of Petroleum
Uttarakhand, India

Kritika Gupta
Amity Institute of Immunology and Virology
Amity University
Uttar Pradesh, India

G. Guru-Pirasanna-Pandi
Crop Protection Division
ICAR-National Rice Research Institute
Cuttack, India

Sarah González Henao
Departamento de Ciencias Biológicas
Universidad ICESI
Valle del Cauca, Colombia

Thaura Ghneim-Herrera
Departamento de Ciencias Biológicas
Universidad ICESI
Valle del Cauca, Colombia

E. Janeeshma
Department of Botany
MES Keveeyam College
Kerala, India

Joy M. Joel
Plant Physiology and Biochemistry Division
Department of Botany
University of Calicut
Kerala, India

Riya Johnson
Plant Physiology and Biochemistry Division
University of Calicut
Kerala, India

Kuldeep Joshi
Center for Biodiversity Conservation and Management
G.B. Pant National Institute of Himalayan Environment
Uttarakhand, India

Khushneet Kaur
Amity Institute of Immunology and Virology
Amity University
Uttar Pradesh, India

Hadjer Kecies
Laboratory of Natural Science and Materials (LSNM)
Institute of Science and Technology
Abdelhafid Boussouf University Centre Mila
Mila, Algeria

Bahman Khoshru
Department of Soil Science, Faculty of Agriculture
University of Tabriz
Tabriz, Iran

Rajesh Kumar
Department of Environmental Microbiology
Babasaheb Bhimrao Ambedkar (A Central) University
Lucknow, India

Abdelkader Mahrouk
Laboratory of Applied Biochemistry
Department of Biochemistry and Molecular and Cellular Biology
Faculty of Nature and Life Sciences
University of Mentouri Constantine
Constantine, Algeria

Mamun Mandal
Laboratory of Applied Stress Biology
Department of Botany
University of Gour Banga
West Bengal, India

Sampriti Mohanty
Crop Protection Division
ICAR-National Rice Research Institute
Cuttack, India

Shuvendu Shekhar Mohapatra
Crop Improvement Division
ICAR-National Rice Research Institute
Cuttack, India
and
Department of Biotechnology
Berhampur University
Berhampur, India

Nutan Moharana
Crop Improvement Division
ICAR-National Rice Research Institute
Cuttack, India
and
Department of Plant Breeding and Genetics
Odisha University of Agriculture and Technology
Bhubaneswar, India

Partha Chandra Mondal
Division of Agricultural Chemicals
Indian Agricultural Research Institute
New Delhi, India

Shokufeh Moradi
Department of Soil Science, Faculty of Agriculture
University of Tabriz
Tabriz, Iran

G. Pagnani
Department of Life, Health and Environmental Sciences
University of Teramo
Teramo, Italy

Muktipada Panda
Department of Botany
Banki College (Auto.)
Odisha, India
and
Department of Higher Education
Government of Odisha
Odisha, India

Sonali Panda
Crop Physiology and Biochemistry Division
ICAR-National Rice Research Institute
Cuttack, India
and
Department of Botany
Ravenshaw University
Cuttack, India

Subhadra Pattanayak
Plant Breeding and Genetics
Regional Research and Technology Transfer Stations
Odisha, India

M. Pellegrini
Department of Life, Health and Environmental Sciences
University of L'Aquila
Palazzo Camponeschi, Italy

Jorge Poveda
Institute for Multidisciplinary Research in Applied Biology (IMAB)
Universidad Pública de Navarra (UPNA)
Pamplona, Spain
and
Centro de Investigação de Montanha (CIMO)
Instituto Politécnico de Bragança
Bragança, Portugal

S. R. Prabhukarthikeyan
Crop Protection Division
ICAR-National Rice Research Institute
Cuttack, India

Biswajit Pramanik
Department of Genetics and Plant Breeding
Palli Siksha Bhavana (Institute of Agriculture)
West Bengal, India

Ankita Priyadarshini
Crop Production Division
ICAR-National Rice Research Institute
Odisha, India

Shampa Purkaystha
Department of Genetics and Plant Breeding and Seed Science and Technology
Centurion University of Technology and Management
Odisha, India

Aishwary Purohit
Environmental Microbiology Research Group
Department of Life Science
Graphic Era (Deemed to be University)
Uttarakhand, India

Anuprita Ray
School of Bioscience and Biotechnology
Lovely Professional University
Punjab, India

Aya Rehouma
Laboratoire de Biotechnologies
Ecole Nationale Supérieure de Biotechnologie
Ville Universitaire Ali Mendjeli
Constantine, Algeria

Swagatika Sahoo
Crop Protection Division
ICAR-National Rice Research Institute
Cuttack, India

Puranjoy Sar
Department of Genetics and Plant Breeding
Palli Siksha Bhavana (Institute of Agriculture)
West Bengal, India

Abhijit Sarkar
Laboratory of Applied Stress Biology
Department of Botany
University of Gour Banga
West Bengal, India

Sucharita Satapathy
Crop Improvement Division
ICAR-National Rice Research Institute
Cuttack, India

Shraddha Bhaskar Sawant
Department of Plant Pathology
Odisha University of Agriculture & Technology
Odisha, India

Ansuman Senapati
Crop Production Division
ICAR-National Rice Research Institute
Cuttack, India

A. M. Shackira
Department of Botany
Sir Syed College
Kerala, India

Swagat Shubhadarshi
Department of Agronomy
Siksha O Anusandhan University
Odisha, India

Shivangi Singh
Amity Institute of Immunology and Virology
Amity University
Uttar Pradesh, India

Somya Sinha
Department of Biotechnology
Graphic Era (Deemed to be) University
Uttarakhand, India

Pooja Thathola
Center for Biodiversity Conservation and Management
G.B. Pant National Institute of Himalayan Environment
Uttarakhand, India

1 History, Diversity, and Community Dynamics of Arbuscular Mycorrhizal Fungi in the Rice Ecosystem

Debashree Dalai
ICAR-National Rice Research Institute

Muktipada Panda
Banki College (Auto.)
Govt. of Odisha

CONTENTS

1.1 INTRODUCTION

In nature, certain fungi form a symbiotic association with plant roots known as "mycorrhiza." The term was derived by combining two Greek words, *mykes* and *rhiza*, which mean fungi and root, respectively. For the very first time, this term was used by A. B. Frank (a plant pathologist) in 1885 to express the mutualistic relationship between these two partners (Altındal and Altındal, 2019). Six types of mycorrhizal association exist with plant root systems: arbuscular mycorrhiza (AM),

DOI: 10.1201/9781003354086-1

ecto-, ericoid, arbutoid, monotropoid, and orchid mycorrhiza (Smith and Read, 1997). Three taxonomic classes of fungi form mycorrhiza: Archaeosporomycetes, Glomeromycetes, and Paraglomeromycetes (Monika et al., 2019). Among them, AM formed by fungi of the class "Glomeromycetes" is very common. The term "arbuscular" is assigned due to the occurrence of fungal "arbuscules" inside cortical cells of most plant roots. The arbuscular mycorrhizal fungi (AMF) are widespread obligate symbiotic soil fungi that use carbon compounds in the form of carbohydrates and sugar produced by the plant hosts for their growth and reproduction. It is well established that currently, 80% (approx.) of terrestrial land plants configured a mutualistic relationship with AMF (Smith and Read, 2008; Brundrett, 2009). The association allows several ecological benefits attributed to mycorrhizae for the host species, the most common of which is the efficient uptake of immobile nutrients (especially P) from the soil (Mitra et al., 2021), and in trade-off, plants provide up to 30% of their photosynthetically fixed carbon to AMF (Roth and Paszkowski, 2017). In this agricultural system, AMF is central to sustainable farming. Thus, proper management and maintenance of AMF will significantly contribute to agricultural sustainability and profitability (Panneerselvam et al., 2019, 2020). Despite the significant value of AMF in sustainable crop production and improvement, the interaction between AMF and monocots, especially, in rice systems is yet to be explored.

1.2 HISTORICAL EVIDENCE

Paleontological evidence shows that the existence of AMF dates back to the appearance of the first land plant. They have a typical evolutionary history and cellular characteristics. AMF belong to a monophyletic group of fungal subphylum "Glomeromycotina" of the phylum "Mucoromycota" (Spatafora et al., 2016), which evolved more than 430 million years ago during the Ordovician period (Simon et al., 1993; Redecker et al., 2000). This long history of co-evolutionary relationship has led many researchers to suggest that AMF had played a major role in the colonization of vascular land plants. Since their discovery, intensive research and investigations have undergone radical transformations to reveal the taxonomic history and systematics of both partners, especially the fungal partner. According to Schübler and Walker (2011), AMF research has an extensive history in terms of evolution, phylogeny, and taxonomy. The evolutionary history of these obligate biotrophs has been categorized into four different evolutionary periods as follows.

1.2.1 The Discovery Period of AMF (1845–1974)

During this initial period, new species were discovered and described on the basis of recognizable asexual spores. The period began with the identification and description of two species of the genus *Glomus* (i.e., *Glomus microcarpus* and *Glomus macrocarpus*) by the Tulasne brothers (Tulasne and Tulasne, 1845) and ended with the classification by Gerdemann and Trappe (1974). The species of the genus *Sclerocystis* were identified based on the formation of spores born on small sporocarps-like structure by Berkeley and Broome (1873). Thus, genera like *Glomus* and *Sclerocystis* were used for classifications before the establishment of the term "Mycorrhiza" by Frank in 1885 (Altındal and Altındal, 2019). Different spore types were described by letter codes A, B, and C (Gerdemann, 1955). In this period, two new genera, i.e., *Acaulospora* and *Gigaspora*, and twelve new species were also described, as well as many taxa were redefined. Additionally, the genus *Glomus*, which was previously combined with Endogone by the Tulasne brothers, was confirmed as a distinct and valid genus.

1.2.2 The Alpha Taxonomy Period of AMF (1975–1989)

In this period, morphological identification, classification, and description of many new genera and families of glomeromycotan fungi were made. A proposal for standardization and description of new species based on phenotypic characters of fungal spores was also made available as the subcellular structures of AMF spores were much more diverse among species. To overcome the difficulty of classification based on spore structure, Walker (1983) proposed a new method to describe species based on distinct "wall types" of AMF spores of glomeromycotan fungi. He also developed a "murograph" that contains a graphical representation of different wall types and groups in a spore, which is considered one of the most important advancements in the field of taxonomy for glomeromycotan fungi, and the terminology is still being used to describe AMF species. During this time, 77 species of glomeromycotan fungi were listed by Trappe (1982), and 6 years later, 126 new species were listed by Schenck and Pérez (1988). Subsequently, different taxonomic keys were developed, such as the dichotomous key (Hall and Fish, 1979), the synoptic key (Trappe, 1982), and keys for classifying groups of species (Koske and Walker, 1985). Toward the end of this period, most morphological characteristics used for taxonomic identification and classification of AMF were evaluated, and some new methods were also approached (Morton, 1988).

1.2.3 The Cladistics Period of AMF (1990–2000)

This period is marked by the approach of molecular biology to the systematics of AMF, especially the glomeromycotan fungi. Morton (1990) used 57 AMF species identified based on 27 phenotypic characteristics to conclude the hypothesis that glomeromycotan fungi contained a monophyletic group determined by the formation of intraradical arbuscules and mutualistic symbiosis. Another major event that occurred was the usage of small subunit rRNA (SSU) gene sequences to establish the evolutionary relationships among taxa within the order "Glomerales." The ancient origin of AM fungi was also first reported by Simon et al. (1993) by measuring the rate of evolution of the SSU gene surveyed against the record of the fossil using the molecular clock-based approach. This period ends with the recognition of two ancestral clades based on rDNA sequences (Redecker et al., 2000).

1.2.4 The Phylogenetic Synthesis Period of AMF (2001-Present Times)

This period is marked by the modern classification and specification of new taxa based on genetic characters (SSU rRNA gene) as well as based on fossil records. The most notable event that occurred during this period was the naming of a new phylum by Schubler et al. (2001) under the kingdom Fungi to cluster all the species of AMF as well as the *Geosiphon pyriforme*, a little well-known fungus, that creates a symbiosis with *Nostoc* (Schubler and Kluge, 2001). For the first time, the presence of a germination shield in glomeromycotan spores of at least 400 million years old from Lower Devonian Rhynie chert was reported by Dotzler et al. (2006). The radical extension of genera and families in Glomeromycota was approached by Oehl et al. (2008) by interpreting molecular data. Morton and Msiska (2010) reported that the sequences of 25S rDNA appeared with a minimum number of bootstraps in the tree of concatenated sequences of 25S rDNA as well as β-tubulin as per the clades proposed by Oehl et al. (2008). To date, Krüger et al. (2011) proposed a phylogenetic backbone that can be used for new species descriptions based on rDNA sequences.

1.3 DIVERSITY

The actual taxonomic diversity of AMF is still unknown. The integration of both morpho- and genetic taxonomy based on rRNA sequence shows that globally around (±) 300 species of AMF are found (Öpik et al., 2013). But by morphological features, only 150–200 species can be distinguished (Santos-Gonzalez et al., 2007). Classification of AMF based on the structure of fungal hyphae/mycelium and β-tubulin had presented them under the class "Glomeromycetes" of the phylum "Glomeromycota." This class consists of five orders (i.e., Archeosporales, Diversisporales, Gigasporales, Glomerales, and Paraglomerales), which altogether constitute 14 families under 29 genera, and 230 AMF species are present (Monika et al., 2019). In the terrestrial ecosystem, most of the AMF belong to six common genera: *Acaulospora*, *Entrophospora*, *Gigaspora*, *Glomus*, *Sclerocystis*, and *Scutellospora.* So far, ~161 species have been reported from Indian habitats, and the most common species among them are *Acaulospora laevis*, *Acaulospora scrobiculata*, *Acaulospora spinosa*, *Claroideoglomu sclaroideum*, *Claroideoglomus etunicatum*, *Funneliformis mosseae*, *Glomus aggregatum*, *Rhizophagus fasciculatus*, and *Rhizophagus intraradices* (Gupta et al., 2014, 2017). The study also reported that the genera *Glomus* outcompete with others both in abundance and in species richness.

The rice ecosystem is very poorly studied, which limits to provide the exact species diversity of rice AMF as the available data represent only at the regional scale. Recently, based on both molecular and morphological characteristics, 11 rice AMF genera have been identified from rainfed rice cultivars across six different regions in Ghana (Sarkodee-Addo et al., 2020; Table 1.1). The predominant genera encountered in this study were *Glomus*, *Rhizophagus*, *Scutellospora*, and *Acaulospora.* Similarly, Martins and Rodrigues (2020) recorded 17 rice AMF

TABLE 1.1
Diversity of Arbuscular Mycorrhizal Fungi in Different Rice Systems

Ecosystem Type	Country	AM Fungal Genera/Species	References
Lowlands	India	*Acaulospora scrobiculata*, *A. delicate*, *A. dilatata*, *A. laevis*, *A. tuberculata*, *A. myriocarpa*, *A. soloidea*, *Funneliformis mosseae*, *Rhizoglomus fasciculatum*, and *Entrophospora nevadensis*	Martins and Rodrigues (2020)
Midlands		*Acaulospora scrobiculata*, *A. bireticulata*, *A. rehmii*, *A. dilatata Gigaspora ramisporophora*, *Claroideoglomus claroideum*, *Entrophospora nevadensis*, *Funneliformis mosseae*, *Gigaspora albida* and *G. decipiens*	
Uplands		*A. scrobiculata*, *A. bireticulata*, *Claroideoglomus claroideum*, *C. etunicatum*, and *Entrophospora nevadensis*	
Rainfed rice cultivars	Ghana	*Acaulospora*, *Ambispora*, *Archaeospora*, *Claroideoglomus*, *Diversispora*, *Gigaspora*, *Glomus*, *Racocetra*, *Redeckera*, *Rhizophagus*, and *Scutellospora*	Sarkodee-Addo et al. (2020)
Rice	India	*Claroideoglomus* and *Glomus*	Panneerselvam et al. (2020)

species sampled from three different rice ecosystem types, i.e., lowlands, midlands, and uplands. The species recognized in this study were from six genera, i.e., *Acaulospora*, *Claroideoglomus*, *Entrophospora*, *Funneliformis*, *Gigaspora*, and *Rhizoglomus* (Table 1.1). Ecosystem diversity also determines changes in the abundance of particular genera and species as *Acaulospora* was abundant in the lowlands, *Gigaspora* was abundant in the midlands, whereas *Claroideoglomus* was dominant in upland rice fields (Martins and Rodrigues, 2020).

1.4 COMMUNITY DYNAMICS

Rice is cultivated in different environments characterized by its hydrology, such as in irrigated, rainfed lowland, upland, and flood-prone areas (Martins and Rodrigues, 2020). The AMF communities are mostly determined by the plant's internal environment and the soil's physicochemical variables. Seasonal changes in environmental factors affect AMF symbiosis in rice roots. The seasonal variation in water level and duration of waterlogging significantly decide species type and abundance of AMF association in the rice root system (Vallino et al., 2014; Sánchez et al., 2015). Generally, AMF readily form associations in well-aerated soil habitats and are rarely under flooded conditions due to unfavorable hypoxia situations (Maiti et al., 1995; Toppo et al., 2012; Martins and Rodrigues, 2020). Despite being obligate aerobes, AMF can survive some extent of waterlogging, whereas their abundance is drastically reduced (FAO, 2018; Martins and

Rodrigues, 2020; Bao et al., 2022). Like aerobic eukaryotes, oxygen (i.e., O_2) availability in the soil atmosphere plays a critical role in shaping the community characteristics of AMF in natural ecosystems (Macek, 2017).

In the case of the rice ecosystem, different opinions exist on community dynamics of AMF colonization at different growth stages or seasons in a year. Fungal colonization varies within and between years and among plants of different species (Sanders and Fitter, 1992; Lugo et al., 2003; Fuchs and Haselwandter, 2004; Li et al., 2005). Within a year, their highest abundance occurs in summer and declines during winter and early spring (DeMars and Boerner, 1995; Kabir et al., 1997; Lugo et al., 2003). As AMF are aerobic fungi, their establishment in rice plants is strongly related to the formation of root aerenchyma that allows them to obtain oxygen (Vallino et al., 2014). A comparison of the intensities of AMF colonization at different growth stages, i.e., seedling, tillering, heading, and ripening, showed that most associations were established during the heading and ripening stages, which again related to the development of cortical aerenchyma in rice roots (Wang et al., 2015). The poor and rare occurrence of AMF at an early stage of plant development (i.e., seedling and tillering stage) is due to the absence of poorly developed root aerenchyma. By performing a greenhouse experiment, Sánchez et al. (2015) found that rice plants were colonized with AMF only after 35 days of seed germination both under non-flooded and flooded conditions. However, not only a negative relationship has been reported between AMF colonization and flooding in rice ecosystems, but a decreased trend in abundance has also been reported in other wetland ecosystems (Wang et al., 2010; Sánchez et al., 2015).

1.5 CONCLUSION

It could be concluded that the colonization of AMF in plants has a long co-evolutionary history. This symbiotic association plays a potential role in increasing crop productivity. Thus, knowledge on the taxonomic diversity of AMF in rice ecosystems will help toward their use for sustainable rice production.

REFERENCES

Altındal, D., and Altındal, N. 2019. Relationship between field crops and mycorrhiza. In: *Mycorrhizosphere and Pedogenesis*, eds. A. Varma, and D. K. Choudhary, pp. 61–72. Springer Nature Singapore Pte Ltd: Singapore.

Bao, X., Zou, J., Zhang, B., Wu, L., Yang, T., and Huang, Q. 2022. Arbuscular mycorrhizal fungi and microbes interaction in rice mycorrhizosphere. *Agronomy* 12:1277.

Berkeley, M. J., and Broome, C. E. 1873. Enumeration of the fungi of Ceylon: Part II. *J Linn Soc Lond Bot* 14:137.

Brundrett, C. M. 2009. Mycorrhizal associations and other means of nutrition of vascular plants: understanding the global diversity of host plants by resolving conflicting information and developing reliable means of diagnosis. *Plant Soil* 320:37–77.

DeMars, B. G., and Boerner, R. E. J. 1995. Mycorrhizal dynamics of three woodland herbs of contrasting phenology along topographic gradients. *Am J Bot* 82:1426–1431.

Dotzler, N., Krings, M., Taylor, T. N., and Agerer, R. 2006. Germination shields in Scutellospora (Glomeromycota: Diversisporales, Gigasporaceae) from the 400 million-year-old Rhynie chert. *Mycol Prog* 5:178–184.

FAO. 2018. Rice Market Monitor. Available online: https://www.fao.org/3/I9243EN/i9243en.pdf (accessed on 15 October 2022).

Fuchs, B., and Haselwandter, K. 2004. Red list plants: colonization by arbuscular mycorrhizal fungi and dark septate endophytes. *Mycorrhiza* 14:277–281.

Gerdemann, J. W. 1955. Relation of a large soil-borne spore to phycomycetous mycorrhizal infections. *Mycologia* 47:429–617.

Gerdemann, J. W., and Trappe, J. M. 1974. The Endogonaceae in the Pacific Northwest. *Mycol Mem* 5:1–76.

Gupta, M. M., Naqvi, N., and Kumar, P. 2017. iAMF–centralized database of arbuscular mycorrhizal distribution, phylogeny and taxonomy. *J Recent Adv Appl Sci* 30(1):1–5.

Gupta, M. M., Naqvi, N. S., and Singh, V. K. 2014. The state of arbuscular mycorrhizal fungal diversity in India: an analysis. *Sydowia* 66:265–288.

Hall, I. R., and Fish, B. J. 1979. A key to the Endogonaceae. *Trans Br Mycol Soc* 73:261–270.

Kabir, Z., O'Halloran, I. P., Fyles, J. W., and Hamel, C. 1997. Seasonal changes of arbuscular mycorrhizal fungi as affected by tillage practices and fertilization: hyphal density and mycorrhizal root colonization. *Plant Soil* 192:285–293.

Koske, R. E., and Walker, C. 1985. Species of Gigaspora (Endogonaceae) with roughened outer walls. *Mycologia* 77:702–720.

Krüger, M., Krüger, C., Walker, C., Stockinger, H., and Schübler, A. 2011. Phylogenetic reference data for systematics and phylotaxonomy of arbuscular mycorrhizal fungi from phylum to species level. *New Phytol* 193(4):970–984.

Li, L. F., Yang, A., and Zhao, Z. W. 2005. Seasonality of arbuscular mycorrhizal symbiosis and dark septate endophytes in a grassland site in southwest China. *FEMS Microbiol Ecol* 54:367–373.

Lugo, M., González Maza, M. E., and Cabello, M. N. 2003. Arbuscular mycorrhizal fungi in a mountain grassland II: seasonal variation of colonization studied, along with its relation to grazing and metabolic host type. *Mycologia* 95:407–415.

Macek, I. 2017. Arbuscular mycorrhizal fungi in hypoxic environments. In: *Mycorrhiza - Function, Diversity, State of the Art*, eds. A. Varma, R. Prasad, and N. Tuteja, pp. 329–348. Springer: Berlin Heidelberg.

Maiti, D., Variar, M., and Saha, J. 1995. Colonization of upland rice by native VAM under direct mono- cropped ecosystem. In: *Recent Advances in Phytopathological Research*, ed. A. K. Roy, and K. K. Sinha, pp. 45–51. MD Publication Ltd: New Delhi.

Martins, W. F. X., and Rodrigues, B. F. 2020. Identification of dominant arbuscular mycorrhizal fungi in different rice ecosystems. *Agric Res* 9(1):46–55.

Mitra, D., Djebaili, R., Pellegrini, M., Mahakur, B., Sarker, A., Chaudhary, P., Khoshru, B., Gallo, M. D., Kitouni, M., Barik, D. P., and Panneerselvam, P. 2021. Arbuscular mycorrhizal symbiosis: plant growth improvement and induction of resistance under stressful conditions. *J Plant Nutr* 44(13):1993–2028.

Monika, D. S., Arya, S. S., Kumar, N., and Kumar, S. 2019. Mycorrhizal fungi: Potential candidate for sustainable agriculture. In: *Mycorrhizosphere and Pedogenesis*, eds. A. Varma, and D. K. Choudhary, pp. 339–353. Springer Nature Singapore Pte Ltd: Singapore.

Morton, J. B. 1988. Taxonomy of VA mycorrhizal fungi: classification, nomenclature, and identification. *Mycotaxon* 32:267–324.

Morton, J. B. 1990. Evolutionary relationships among arbuscular mycorrhizal fungi in the Endogonaceae. *Mycologia* 82:192–207.

Morton, J. B., and Msiska, Z. 2010. Phylogenies from genetic and morphological characters do not support a revision of Gigasporaceae (Glomeromycota) into four families and five genera. *Mycorrhiza* 20:483–496.

Oehl, F., de Souza, F. A., and Sieverding, E. 2008. Revision of Scutellospora and description of five new genera and three new families in the arbuscular mycorrhizal-forming Glomeromycetes. *Mycotaxon* 106:311–360.

Öpik, M., Davison, J., Moora, M., and Zobel, M. 2013. DNA-based detection and identification of Glomeromycota: the virtual taxonomy of environmental sequences. *Botany* 92(2):135–147.

Panneerselvam, P., Kumar, U., Senapati, A., Parameswaran, C., Anandan, A., Kumar, A., Jahan, A., Padhy, S. R., and Nayak, A. K. 2020. Influence of elevated CO_2 on arbuscular mycorrhizal fungal community elucidated using Illumina MiSeq platform in sub-humid tropical paddy soil. *Appl Soil Ecol* 145:103344.

Panneerselvam, P., Sahoo, S., Senapati, A., Kumar, U., Mitra, D., Parameswaran, C., Anandan, A., Kumar, A., Jahan, A., and Nayak, A. K. 2019. Understanding interaction effect of arbuscular mycorrhizal fungi in rice under elevated carbon dioxide conditions. *J Basic Microbiol* 59(12):1217–1228.

Redecker, D., Kodner, R., and Graham, L. E. 2000. Glomalean fungi from the Ordovician. *Science* 289:1920–1921.

Roth, R., and Paszkowski, U. 2017. Plant carbon nourishment of arbuscular mycorrhizal fungi. *Curr Opin Plant Biol* 39:50–56.

Sánchez, M. R., Baños, Y. S., Hernández, Y. M., Martínez, A. Y., Benitez, M., Bharat, B. V., and Chávez, Y. P. 2015. Arbuscular mycorrhizal symbiosis in rice (*Oryza sativa* L.) plants in flooded and non-flooded conditions. *Acta Agron* 64(3):211–217.

Sanders, I. R., and Fitter, A. H. 1992. The ecology and functioning of vesicular-arbuscular mycorrhizas in co-existing grassland species. Seasonal patterns of mycorrhizal occurrence and morphology. *New Phytol* 120:517–524.

Santos-Gonzalez, J. C., Finlay, R. D., and Tehler, A. 2007. Seasonal dynamics of arbuscular mycorrhizal root colonization in a semi natural grassland. *Appl Environ Microbiol* 73:5613–5623.

Sarkodee-Addo, E., Yasuda, M., Lee, C. G., Kanasugi, M., Fujii, Y., Omari, R. A., Abebrese, S. O., and Bam, R., Asuming-Brempong S., Dastogeer, K. M. G., and Okazaki, S. 2020. Arbuscular Mycorrhizal Fungi Associated with Rice (*Oryza sativa* L.) in Ghana: effect of regional locations and soil factors on diversity and community assembly. *Agronomy* 10:559.

Schenck, N. C., and Pérez, Y. 1988. Manual for the identification of VA mycorrhizal fungi. University of Florida, Gainesville.

Schubler, A., and Kluge, M. 2001. Geosiphonpyriforme, an endocytosymbiosis between fungus and cyanobacteria, and its meaning as a model system for arbuscular mycorrhizal research. In: *The Mycota*, ed. B. Hock, pp. 151–161. Springer: Berlin Heidelberg, New York.

Schübler, A., and Walker, C. 2011. Evolution of the ‘plant-symbiotic’ fungal phylum, glomeromycota. In: *Evolution of Fungi and Fungal-Like Organisms, the Mycota XIV*, eds. S. Poggeler, and J. Wostemeyer, pp. 163–185. Springer Verlag: Berlin Heidelberg.

Schubler, A., Schwarzott, D., and Walker, C. 2001. A new fungal phylum, the Glomeromycota: phylogeny and evolution. *Mycol Res* 105:1413–1421.

Simon, L., Bousquet, J., Levesque, R. C., and Lalonde, M. 1993. Origin and diversification of endomycorrhizal fungi and coincidence with vascular land plants. *Nature* 363:67–69.

Smith, S. E., and Read, D. J. 1997. *Mycorrhizal Symbiosis*. Academic Press: New York.

Smith, S. E., and Read, D. J. 2008. *Mycorrhizal Symbiosis*. Academic Press: London.

Spatafora, J. W., Chang, Y., Benny, G. L., Lazarus, K., Smith, M. E., Berbee, M. L., et al. 2016. A phylum-level phylogenetic classification of zygomycete fungi based on genome-scale data. *Mycologia* 108:1028–1046.

Toppo, N. N., Maiti, D., and Srivastava, A. K. 2012. Native arbuscular mycorrhizal fungal diversity in rice- based cropping system under rainfed ecosystem. *Columban J Life Sci* 13(1&2):79–85.

Trappe, J. M. 1982. Synoptic key to the genera and species of zygomycetous mycorrhizal fungi. *Phytopathol* 72:1102–1108.

Tulasne, L. R., and Tulasne, C. 1845. Fungi nonnulli hypogaei, novi v. minus cogniti act. *Giorn Bot Ital* 2(Pt. 1):35–63.

Vallino, M., Fiorilli, V., and Bonfante, P. 2014. Rice flooding negatively impacts root branching and arbuscular mycorrhizal colonization, but not fungal viability. *Plant Cell Environ* 37:557–572.

Walker, C. 1983. Taxonomic concepts in the Endogonaceae: spore wall characteristics in species descriptions. *Mycotaxon* 18:443–455.

Wang, Y., Li, T., Li, Y., Björn, L. O., Rosendahl, S., Olsson, P. A., Li, S., and Fu, X. 2015. Community dynamics of arbuscular mycorrhizal fungi in high-input and intensively irrigated rice cultivation systems. *Appl Environ Microbiol* 81:2958–2965.

Wang, Y. T., Qui, Q., Yang, Z. Y., Hu, Z. J., Fung-Yee, N. T., and Xin, G. R. 2010. Arbuscular mycorrhizal fungi in two mangroves in South China. *Plant Soil* 33:181–191.

2 Metagenomics to Explore Mycorrhizal Diversity in Rice Ecosystem

Shokufeh Moradi and Bahman Khoshru
University of Tabriz

Debasis Mitra
Raiganj University

CONTENTS

2.1 INTRODUCTION

One of the most important agricultural products in the world is rice (*Oryza sativa* L.). It has been confirmed that more than 50% of the world's population feeds on this agricultural product. The increase in the growing world population has increased the demand for this product (FAO, 2021).

There are many microorganisms living in the rice plant rhizosphere that interact with its roots. These microbes play an important role in the health of the rice crop (Ding et al., 2019). Arbuscular mycorrhizal fungi (AMF) belonging to the Glomeromycota are one of the important symbiotic microbes of the rice rhizosphere, which provide a range of soil nutrients (such as P and N) to plants in exchange for plant carbohydrates (Vallino et al., 2014; Bao et al., 2019). AMF can improve the biotic and abiotic resistance of rice plants. The extensive hyphae produced by AMF in rice soil are a habitat for microbes in the rhizosphere and microbes in the cytoplasm of some fungal species (Desiro et al., 2014; Venice et al., 2020). Although there are reports that AMF is not present in rice roots

DOI: 10.1201/9781003354086-2

of flooded paddy fields (Ilag et al., 1987; Lumini et al., 2011; Panneerselvam et al.; 2017), many studies have AMF colonization in rice roots in paddy fields (Chen et al., 2017; Bernaola et al., 2018; Sahoo et al., 2017). The response of rice plants to AMF varies from positive to negative (Bao et al., 2019; Lin et al., 2014; Wang et al., 2021; Mitra et al., 2021a,b). Identifying microbes living in the rice rhizosphere is a great challenge in agricultural science due to its high microbial diversity and complex and variable matrix (Tiedje et al., 1999).

AMF cannot be cultivated and propagated independently in the laboratory (*in vitro*), and only few AMF, e.g., AMF that belong to the *Rhizoglomus* genus, can grow and reproduce under these conditions (Declerck et al., 2005). The lack of successful cultivation of AMF axenically in the laboratory strengthens the hypothesis that they are meta-organisms inseparable from their bacterial and fungal symbionts. Whole genome sequencing (WGS) has been performed for AMF species that can grow and reproduce under *in vitro* conditions. Although this provides important genetic information, there are limitations in serving as reference genomes due to the high genomic variability between and within AMF isolates (Tisserant et al., 2013; Boon et al., 2015). Because most AMF cannot be cultured under *in vitro* conditions, WGS AMF data can provide valuable information for relevant researchers. Also, whole metagenome sequencing (WMS) data studies can provide valuable information for research related to AMF genomics and the interaction of AMF with other microbes (Kang et al., 2020).

For the assessment and discovery of new species, metagenomics is very useful, although AMF studies with this technique have limitations. Studying the life cycle of AMF revealed that these microbes spend only a part of their life freely, and in most cases, they form symbiosis with crops, forest trees, and wild grasses (indeed over 90% of plant species) (Ganeshamurthy et al., 2017). In this symbiosis, both parties benefit from the relationship, the fungus improves water absorption and nutrient status of the host plants and increases their resistance to diseases, and in return, the host plant benefits the fungus from its photosynthetic products and organic compounds. Since most of the microbes have not yet been identified, and in the case of AMF, maintaining obligate symbiotic fungi in the laboratory is difficult (and sometimes impossible), the need for approaches such as metagenomics to record genome collections of different and new species is felt more and more (Bhargava et al., 2019).

The aim of this chapter is to focus on AMF to provide a short summary of their main biological features and an update on the techniques applied to investigate the biodiversity of this still enigmatic group of soil fungi in the rice rhizosphere.

2.2 METAGENOMICS AND THE AMF COMMUNITY

Metagenomics tries to identify and discover all members of a community, regardless of their number and cultivability in the metagenome. Overcoming this challenge is the goal of metagenomic strategies (Bhargava et al., 2019; Thomas et al., 2012; Panneerselvam et al., 2020). In metagenomics, two main methods are used depending on the purpose of the study. In the first method (targeted metagenomics), DNA isolated from the environment is used as a template and then its amplification is performed using primers against known microorganisms (Bhargava et al., 2019). The PCR product is identified using molecular markers. Therefore, in this technique, the composition of a specific soil community is identified by amplifying a target region of template DNA (extracted from mycorrhizal roots).

In the following, amplicons are cloned using appropriate vectors and hosts so that individual fragments can be identified by sequencing. The clone library analysis technique has led to the identification of many arbuscular mycorrhizae (AM) fungal taxa that other methods (spore techniques) have not been able to identify. The characteristics of sequence-based metagenomics are collecting a large amount of information for a microbial community under investigation, reducing sequencing costs and shorter analysis time, which has revolted quantitative metagenomics. So, it has become the main strategy to study and discover the diversity of AMF.

The second approach is shotgun metagenomics, which is based on random sequencing. In this method, the entire metagenome can be evaluated, and therefore, it is possible to study the entire gene content as well as the entire community structure. In metagenomics that is based on function, without having information about the gene sequence, protein structure, or microbe of origin, it is possible to access the genetic diversity of each community. In this approach, an enormous number of random DNA fragments are translated into proteins through bacteria that grow and reproduce in the laboratory. "Foreign" proteins produced after gene expression by clones are screened in different ways such as antibiotic resistance or vitamin production (Bhargava et al., 2019; Thomas et al., 2012; Mitra et al., 2022). Figure 2.1 shows a summary of metagenomics steps.

2.3 GENOMIC DNA EXTRACTION AND PURIFICATION OF RICE RHIZOSPHERE

DNA extracted (direct and indirect extraction) from a microbial population such as AMF must be representative of all members of the metagenome, and this step is a significant challenge (Delmont et al., 2011). Extraction should ideally be performed and include an unbiased analysis of the source DNA. Cell lysis should be performed in such a way that it does not damage the DNA of other members of the community, and in drawing conclusions about the overall metabolic capabilities of a microbial community, it is important to distinguish between the DNA of living and dead cells in a given sample (Marotz et al., 2017; Delmont et al., 2011; Wesolowska-Andersen et al., 2014).

2.3.1 Indirect Method

In this method, DNA is extracted from spores, which are actually the dormant phase of the fungus life cycle. Taxonomic expertise is needed to isolate spores from soils, and morphological identification is of great importance in most AMF

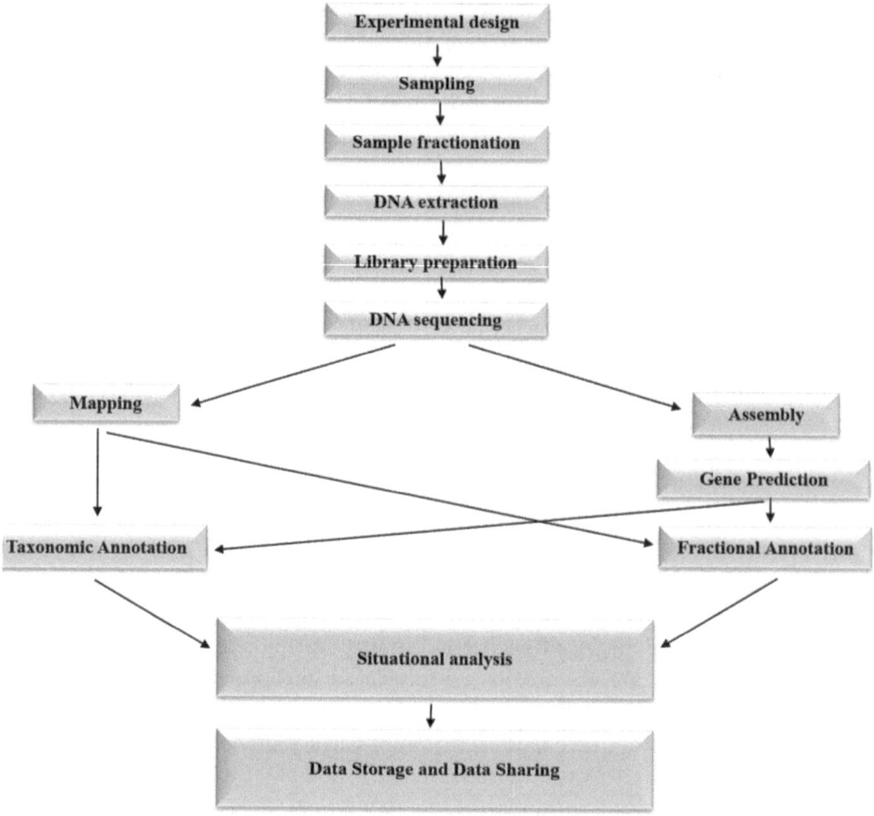

FIGURE 2.1 Flow diagram of typical metagenome projects.

studies that investigate diversity (Gai et al., 2006). In this strategy, soil samples (trap cultures) are used as an inoculum on pot cultures to see the proliferation of local AMF species (Mathimaran et al., 2005; Panneerselvam et al., 2019). In this method, there are some criticisms that only one stage of the microbe's life is investigated and the assessment of spore density in field samples may not correspond to the AMF population that actually colonizes the roots. Additionally, it is impossible to distinguish between the spores produced in the current season and other seasons, and it is also possible that some species do not produce spores all the time. It has been reported that most spores among fungal species are related to *Glomus* species, and the type of farming determines the abundance and distribution of spores in the soil (Delmont et al., 2011).

2.3.2 Direct Method

The progress of molecular techniques in the study and evaluation of fungal diversity in the field of phylogenetics has led to many revolutions in fungal phylogenetics and ecology. To remove unwanted RNA, the addition of RNase-A is used in the purification of metagenomic DNA (from soil and roots) extracted by kits (manual and commercial). The presence of impurities or contaminants (even minor) could have a negative effect on the recovery of DNA or RNA and could reduce the efficiency of sequencing, thus increasing the cost of the experiment and decreasing the chance of recovery for the community with a less number. For the initial assessment of any microbial community such as AMF, PCR amplification is necessary to describe the cloning and sequencing of that microbial community with the help of hybridizable primers with highly conserved regions in genes (Panneerselvam et al., 2020). With the help of this methodology, very comprehensive information has been obtained about the structure, composition, diversity, and species richness of different microbial communities (Mathimaran et al., 2005; Delmont et al., 2011).

2.4 CONCLUSIONS AND FUTURE PROSPECTS

Among the different types of mycorrhizae, AM are considered the ancestral and dominant type. Rice, as a strategic plant, has various interactions with microbes in its rhizosphere, and one of the most important interactions is plant–AMF interactions, which is the most influential process on the growth and performance of rice, as well as on soil structure. Since only a small percentage (0.1%–1%) of soil microbes can be cultured in the laboratory with the conventional culture method, most diversity of soil microbial communities is unknown, and only a small part of the gene pool has been isolated and identified with this method. Metagenomics is a new approach to study and identify uncultivable microbes. The methodological strategy used and the computational tools chosen for sequence analysis determine the output of metagenomic studies. This is why bioinformatics has quickly become one of the main challenges and bottlenecks in metagenomic research. Combining bioinformatics and barcoding with metagenomics can speed up the assessment and discovery of mycorrhizal fungal communities; however, there is still room for significant development in this field.

REFERENCES

Bao, X., Wang, Y., & Olsson, P. A. (2019). Arbuscular mycorrhiza under water: Carbon–phosphorus exchange between rice and arbuscular mycorrhizal fungi under different flooding regimes. *Soil Biology and Biochemistry*, 129, 169–177.

Bernaola, L., Cange, G., Way, M. O., Gore, J., Hardke, J., & Stout, M. (2018). Natural colonization of rice by arbuscular mycorrhizal fungi in different production areas. *Rice Science*, 25(3), 169–174.

Bhargava, P., Vats, S., & Gupta, N. (2019). Metagenomics as a tool to explore Mycorrhizal fungal communities. In: A. Varma, & D. K. Choudhary (Eds.), *Mycorrhizosphere and Pedogenesis* (pp. 207–219). Springer: Singapore.

Boon, E., Halary, S., Bapteste, E., & Hijri, M. (2015). Studying genome heterogeneity within the arbuscular mycorrhizal fungal cytoplasm. *Genome Biology and Evolution*, 7(2), 505–521.

Chen, X. W., Wu, F. Y., Li, H., Chan, W. F., Wu, S. C., & Wong, M. H. (2017). Mycorrhizal colonization status of lowland rice (*Oryza sativa* L.) in the southeastern region of China. *Environmental Science and Pollution Research*, 24(6), 5268–5276.

Declerck, S., Séguin, S., & Dalpé, Y. (2005). The monoxenic culture of arbuscular mycorrhizal fungi as a tool for germplasm collections. In: S. Declerck, D. G. Strullu, & A. Fortin (Eds.), *In Vitro Culture of Mycorrhizas* (pp. 17–30). Springer: Berlin, Heidelberg.

Delmont, T. O., Robe, P., Clark, I., Simonet, P., & Vogel, T. M. (2011). Metagenomic comparison of direct and indirect soil DNA extraction approaches. *Journal of Microbiological Methods*, 86(3), 397–400.

Desiro, A., Salvioli, A., Ngonkeu, E. L., Mondo, S. J., Epis, S., Faccio, A., … & Bonfante, P. (2014). Detection of a novel intracellular microbiome hosted in arbuscular mycorrhizal fungi. *The ISME Journal*, 8(2), 257–270.

Ding, L. J., Cui, H. L., Nie, S. A., Long, X. E., Duan, G. L., & Zhu, Y. G. (2019). Microbiomes inhabiting rice roots and rhizosphere. *FEMS Microbiology Ecology*, 95(5), fiz040.

FAO. (2022). World food and agriculture—statistical yearbook 2021. *World Food and Agriculture-Statistical Yearbook*.

Gai, J. P., Christie, P., Feng, G., & Li, X. L. (2006). Twenty years of research on community composition and species distribution of arbuscular mycorrhizal fungi in China: a review. *Mycorrhiza*, 16(4), 229–239.

Ganeshamurthy, A. N., Sharma, K., Mitra, D., Radha, T. K., & Rupa, T. R. (2017). Isolation and characterization of arbuscular mycorrhizal fungi and their role in plants growing under harsh environments. *Mycorrhiza News (TERI Press)* 29(3), 7–12.

Ilag, L. L., Rosales, A. M., Elazegui, F. A., & Mew, T. W. (1987). Changes in the population of infective endomycorrhizal fungi in a rice-based cropping system. *Plant and Soil*, 103(1), 67–73.

Kang, J. E., Ciampi, A., & Hijri, M. (2020). SeSaMe PS function: Functional analysis of the whole metagenome sequencing data of the arbuscular mycorrhizal fungi. *Genomics, Proteomics & Bioinformatics*, 18(5), 613–623.

Lin, A., Zhang, X., & Yang, X. (2014). Glomus mosseae enhances root growth and Cu and Pb acquisition of upland rice (*Oryza sativa* L.) in contaminated soils. *Ecotoxicology*, 23(10), 2053–2061.

Lumini, E., Vallino, M., Alguacil, M. M., Romani, M., & Bianciotto, V. (2011). Different farming and water regimes in Italian rice fields affect arbuscular mycorrhizal fungal soil communities. *Ecological Applications*, 21(5), 1696–1707.

Marotz, C., Amir, A., Humphrey, G., Gaffney, J., Gogul, G., & Knight, R. (2017). DNA extraction for streamlined metagenomics of diverse environmental samples. *Biotechniques*, 62(6), 290–293.

Mathimaran, N., Ruh, R., Vullioud, P., Frossard, E., & Jansa, J. (2005). Glomus intraradices dominates arbuscular mycorrhizal communities in a heavy textured agricultural soil. *Mycorrhiza*, 16(1), 61–66.

Mitra, D., Be, G. S., Khoshru, B., De Los Santos Villalobos, S., Belz, C., Chaudhary, P., Shahri, F. N., Djebaili, R., Adeyemi, N. O., El-Ballat, E. M., & El-Esawi, M. A. (2021a). Impacts of arbuscular mycorrhizal fungi on rice growth, development, and stress management with a particular emphasis on strigolactone effects on root development. *Communications in Soil Science and Plant Analysis*, 52(14), 1591–1621.

Mitra, D., Djebaili, R., Pellegrini, M., Mahakur, B., Sarker, A., Chaudhary, P., Khoshru, B., Gallo, M. D., Kitouni, M., Barik, D. P., & Panneerselvam, P. (2021b). Arbuscular mycorrhizal symbiosis: plant growth improvement and induction of resistance under stressful conditions. *Journal of Plant Nutrition*, 44(13), 1993–2028.

Mitra, D., Saritha, B., Janeeshma, E., Gusain, P., Khoshru, B., Nouh, F. A. A., Rani, A., Olatunbosun, A. N., Ruparelia, J., Rabari, A., & Mosquera-Sánchez, L. P. (2022). Arbuscular mycorrhizal fungal association boosted the arsenic resistance in crops with special responsiveness to rice plant. *Environmental and Experimental Botany*, 193, 104681.

Panneerselvam, P., Kumar, U., Senapati, A., Parameswaran, C., Anandan, A., Kumar, A., Jahan, A., Padhy, S. R., & Nayak, A. K. (2020). Influence of elevated CO_2 on arbuscular mycorrhizal fungal community elucidated using Illumina MiSeq platform in sub-humid tropical paddy soil. *Applied Soil Ecology*, 145, 103344.

Panneerselvam, P., Kumar, U., Sugitha, T. C. K., Parameswaran, C., Sahoo, S., Binodh, A. K., Jahan, A., & Anandan, A. (2017). Arbuscular mycorrhizal fungi (AMF) for sustainable rice production. In: T. K. Adhya, B. Lal, B. Mohapatra, D. Paul, & S. Das (Eds.), *Advances in Soil Microbiology: Recent Trends and Future Prospects* (pp. 99–126). Springer: Singapore.

Panneerselvam, P., Sahoo, S., Senapati, A., Kumar, U., Mitra, D., Parameswaran, C., Anandan, A., Kumar, A., Jahan, A., & Nayak, A. K. (2019). Understanding interaction effect of arbuscular mycorrhizal fungi in rice under elevated carbon dioxide conditions. *Journal of Basic Microbiology*, 59(12), 1217–1228.

Sahoo, S., Panneerselvam, P., Chowdhury, T., Kumar, A., Kumar, U., Jahan, A., Senapati, A., & Anandan, A. (2017). Understanding the AM fungal association in flooded rice under elevated CO_2 condition. *ORYZA-An International Journal on Rice*, 54(3), 290–297.

Thomas, T., Gilbert, J., & Meyer, F. (2012). Metagenomics-a guide from sampling to data analysis. *Microbial Informatics and Experimentation*, 2(1), 1–12.

Tiedje, J. M., Asuming-Brempong, S., Nüsslein, K., Marsh, T. L., & Flynn, S. J. (1999). Opening the black box of soil microbial diversity. *Applied Soil Ecology*, 13(2), 109–122.

Tisserant, E., Malbreil, M., Kuo, A., Kohler, A., Symeonidi, A., Balestrini, R.,... & Martin, F. (2013). Genome of an arbuscular mycorrhizal fungus provides insight into the oldest plant symbiosis. *Proceedings of the National Academy of Sciences*, 110(50), 20117–20122.

Vallino, M., Fiorilli, V., & Bonfante, P. (2014). Rice flooding negatively impacts root branching and arbuscular mycorrhizal colonization, but not fungal viability. *Plant, Cell & Environment*, 37(3), 557–572.

Venice, F., Ghignone, S., Salvioli di Fossalunga, A., Amselem, J., Novero, M., Xianan, X.,... & Bonfante, P. (2020). At the nexus of three kingdoms: the genome of the mycorrhizal fungus Gigaspora margarita provides insights into plant, endobacterial and fungal interactions. *Environmental Microbiology*, 22(1), 122–141.

Wang, Y., Bao, X., & Li, S. (2021). Effects of arbuscular mycorrhizal fungi on rice growth under different flooding and shading regimes. *Frontiers in Microbiology*, 12, 756752.

Wesolowska-Andersen, A., Bahl, M. I., Carvalho, V., Kristiansen, K., Sicheritz-Pontén, T., Gupta, R., & Licht, T. R. (2014). Choice of bacterial DNA extraction method from fecal material influences community structure as evaluated by metagenomic analysis. *Microbiome*, 2(1), 1–1.

3 Arbuscular Mycorrhizal Fungi

For Nutrient Management in Rice

Anuprita Ray
Lovely Professional University

Shuvendu Shekhar Mohapatra
ICAR-National Rice Research Institute
Berhampur University

CONTENTS

3.1 INTRODUCTION

Arbuscular mycorrhizal fungi (AMF) are soilborne microorganisms that are fostered as biofertilizers over years, yet their impending abilities to develop the nutritional qualities of crops have not been addressed (Ryan and Graham, 2002; Smith et al., 2009; Panneerselvam and Saritha, 2017). But a sudden thrust for an increase in sustainable agriculture and enhanced qualitative nutritional benefits of crops has reawakened the significance of AM fungal research. AMF have coerced symbionts, which fit into the phylum Glomeromycota, and they outline reciprocal symbioses with more than 80% of cultivated crops (AMF as natural biofertilizers). They establish a symbiotic association with the host and supply water and minerals to plant cells, and in return, they uptake photosynthetic products (Smith and Read, 2008; Poovarasan et al., 2013). According to Smith and Smith (2012) in a regular biological

DOI: 10.1201/9781003354086-3

environment, a non-mycorrhizal circumstance can be termed as uncharacteristic for most of the species. There is a huge variation among AM fungal microflora underneath in the soil condition, and miscellany varies with different plant species, soil types, and environmental factors (Smith and Smith, 2012). Apart from being a great source of nutrients for plants, AM confers other advantages to plants like conferring abiotic stress, i.e., drought and salinity tolerance (Augé et al., 2015) and biotic stress and disease resistance (Pozo and Azcón-Aguilar, 2007; Mitra et al., 2021b). AMF can be called the "need of the hour." In erratic climatic conditions such as soil erosion, irregular rainfall patterns, and the prevalence of extreme climatic conditions, AMF can prove to be a boon to the ecosystem (Panneerselvam et al., 2017, 2019). AMF have an added advantage and are already established as alleviators of heavy metal toxicity in plants and are expected to withstand high metal concentrations in the soil (Meier et al., 2015; Mitra et al., 2021a, c). According to (Bender et al., 2014), AMF can alleviate greenhouse gas emissions by reducing N_2O emissions by increasing plant N absorption and fixation.

Current advances in research suggest that the occurrence of AM in the terrestrial plant system had its own importance, and AM can be termed as a monophyletic creation that has pushed further quick colonization of vascular plants (Delaux, 2017; Sahoo et al., 2017). AM fungal symbiosis with plant species is so prevalent today that most plant species in most ecological habitats (not applicable for hydrophytes) are involved in a symbiotic association.

AMF can address various climatic constraints starting from abiotic to biotic, and the vision ends with lesser greenhouse gas emissions (Panneerselvam et al., 2019). They can address multiple issues and improvise the ecological conditions at one call. However, the utilization of these fungi in various applied programs requires background knowledge about AMF adaptation and responses to target ecological conditions for effective soil management strategies (Dash & Mohapatra, 2018).

3.2 RECENT ADVANCES IN THE NUTRIENT MANAGEMENT ABILITIES OF AMF

The identification of high-effective phosphate transporter (PT) in an AM fungus, the phosphorous uptake by AMF, and the nutritional benefits of AM symbiosis have been a thrust area of research, and both physiological and molecular characterizations have been premeditated extensively (Smith and Smith, 2012). AMF can enhance the nutrient absorption capacity of plants from soil, and they can possess a symbiotic Pi endorsement pathway. Mycorrhiza-inducible ammonium transporters (AMT) are also recently identified transporters with the potential for nutritional enrichment (Kobae et al., 2010). The plant-derived membrane, dubbed as peri-arbuscular membrane, which envelopes the arbuscular, is identified as the site for mineral/nutrient transfer. These membranes have plant transporters that can absorb minerals/nutrients from the peri-arbuscular apoplast and carry these forward to the cortical cells. Harrison et al. (2002) reported that AMT are situated in the peri-arbuscular membranes of Medicago and Soybean and are associated with the Pi transporters (MtPT4) in *Medicago truncatula*. Accordingly, the studies on Pi and ammonium transporter *Medicago* mutants have verified that the Pi and AMT symbiotic transporters (PT4 and AMT 2:3) impact the arbuscular life cycle (Breuillin-Sessoms et al., 2015). This study also demonstrated that the transport of Pi or ammonium through these transporters not only carries nutrients to the root cells but also activates signaling that empowers conditions for arbuscule upkeep. Consequently, AMF are a potential sustainable tool to improve the concentration of micronutrients in plants as an alternative to or as an addition to biofortification using agronomic or genetic approaches. AMF also play an important role in plant nutrition and have an effective impact on crop quality by enriching both micro- and macronutrients through the process of symbiosis. From an agricultural perspective, it may be valuable for multifunctionality quality in nutrient uptake, environmentally friendly, and enhance crop quality.

3.3 ROLE OF AMF IN SOIL PHYSICAL PROPERTIES

AMF have a beneficial effect on soil structure. The arbuscular mycelium is present in huge quantities on the soil surface, ranging from about 80 to 110 m/cm^3 of soil area. Creating stable soil aggregation is the basic property of these mycelia or hyphae. The extra metrical mycelium produces glomalin (glycoprotein) making the AMF a stable, reliable, and long-term soil-binding agent in agricultural practices (Azcon et al., 2013). This glomalin is having several stress tolerant properties like high temperature and hydrophobic conditions of the soil. The AMF have the potential not only to uplift the bioremediation ability but also to protect or maintain soil health through various processes (Figure 3.1). They act like the glue that aggregates soil by togethering clay, silt,

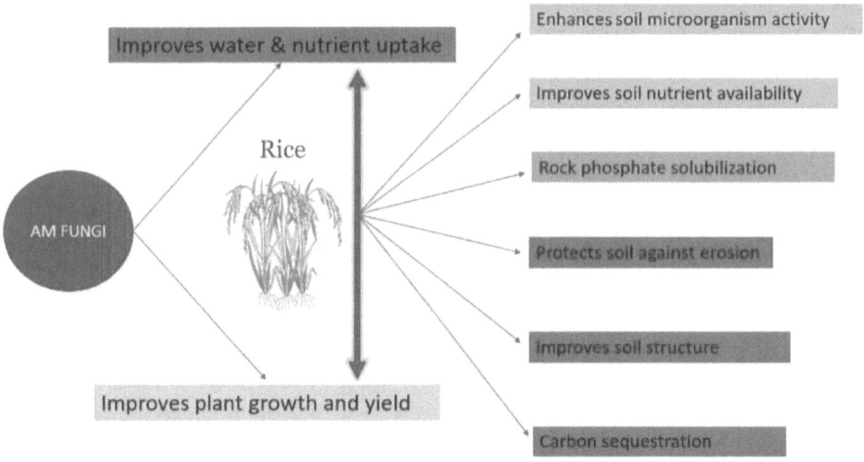

FIGURE 3.1 Functional properties of AMF in a rice cropping system.

and fine sand materials in the soil. It can be assumed that AMF improve soil fertility through their biophysical and chemical mechanisms. The AMF symbionts have been recognized as prime microbial components in soil, which maintain the nutrients (P, N, and C) in adequate concentrations. The AMF appear to be one of the major effective soil microorganisms to consider for being involved in plant mineral nutrition, the ability of water absorption, and protection against several abiotic and biotic stress. Despite the myth that the significance of AMF toward improving soil fertility is well stabilized, our prime understanding of the underlying mechanisms or facts is still limited. There are only few studies have been performed to investigate the simultaneous effect or importance of AMF in enhancing the physical, chemical, and biological aspects (Prasad et al., 2005).

Further research must understand the soil phosphorous and nitrogen critical thresholds below which AM fungi establish symbiosis, soil type, and environmental parameters. In this way, AM fungi serve as a friend to the soil.

3.4 IDENTIFICATION OF AMF-SPECIFIC PHOSPHATE TRANSPORTERS

Molecular genetic research on AMF symbiosis in rice emphasizes the identification of several transporters that are associated with AMF-specific phosphate transportation. Paszkowski et al. (2002) reported that 13 phosphate transporters were identified and that the *OsPT11* transporter is specific to AMF symbiosis in crop plants. The identified transporter (*OsPT11*-GFP) combination protein is precisely localized in the peri-arbuscular membrane adjacent to young and mature arbuscules of AMF, where active phosphate transfer appears to transpire (Kobae et al., 2010). By undergoing *OsPT11* and *OsPT13* gene knockout and knockdown mutants, Yang et al. (2012) have revealed that not only *OsPT11* but also *OsPT13* is intricate in AM symbiosis development in the roots of crop plants. Although *OsPT13* is important for AM symbiosis, symbiotic phosphate uptake is self-regulating of *OsPT13* transporter. These consequences designate purposeful specialization: *OsPT11* may be accountable for both AMF and crop plants.

3.5 INVOLVEMENT OF AMF IN SOIL NITROGEN AVAILABILITY

Nitrogen (N) is also a vital nutrient for the plant just like phosphorus. It generally consists of amino acids, co-enzymes, and phospholipids. Nitrogen is present in forms like nitrites, nitrates, and ammonium ions (organic and mineral components) on the soil surface. Plants generally prefer nitrogen in the form of nitrates (NO_3^-), because ammonium ions are weakly absorbed by plant roots. In the soil, AMF were applied to mobilize the inorganic forms of nitrogen, i.e., NH_4^+. The AMF mycelium is capable of absorbing the ammonium ions (NH_4^+) in the form of nitrates (NO_3^-) and amino acids from the soil and then transporting them to the root systems of crop plants. Definitely, AMF progress the deprivation of organic matter to upsurge nitrogen bioavailability on the soil surface. Nitrogen availability in soil requires the action of resident transporters in the AMF hyphae/mycelium (Ma et al., 2008). It has also been established that mycorrhizal associations could play a substantial role in the decomposition and mineralization of plant organic matter and help mobilize nutrients, predominantly nitrogen, for the advantage of the host crop plant.

3.6 INVOLVEMENT OF AMF IN SOIL-BIOLOGICAL PROPERTIES

Several microorganisms are present in soil and serve as a major constituent of soil structure. These several microorganisms interrelate with each other and with their ecosystem to participate in the operation of the soil and thus contribute to the provision of environmental services essential for our survival using plant production, detoxification of many heavy metal pollutants, and others. The AM fungal diversity is the prime component of soil biota that can determine the biological productive capacity of the ecosystem. These fungal associations vary across crop plants and adjoint ecosystems, but the chief ecological processes that nurture the structure of these diversities are still largely unrevealed. A better understanding of these soil microbiota and their ecological factors can shape their mode of action and affects much more functioning of ecosystems. Soil is then a very energetic biological reactor where various biochemical reactions and essential ecological progressions happen (in specific, the breakdown of organic matter and the biogeochemical cycles of the elements). The microbial activities in the soil contribute to its fertility through synergies between the microbiota, struggle, and parasitism (Igiri et al., 2021).

3.7 ADVANTAGES OF AMF FOR SUSTAINABLE AGRICULTURE

In sustainable agricultural practices, the indebtedness of the essential importance of quality soil life is growing gradually through plant associations. Among these symbiotic microbes, mycorrhiza is one of the important players, which works on the colonization of plant roots. Several studies have been performed to understand the mycorrhizal ecosystem, in particular AMF (Jena et al., 2018). The AMF, by mutual symbiosis, play several functional roles in plant growth, eradication of heavy metal stress, and enhancement of soil fertility, leading to sustainable agricultural practices. AMF hyphae/mycelium and root litter are the furthermost copious carbon source in the soil structure (Gosling et al., 2006). Consequently, AMF deliver an improved supply of energy for soil microbiota to embellish. The datum that AMF affect plant groups is also considered one of the possible mechanisms by which AMF impact soil microbial diversity (Rahman et al., 2020).

3.8 CONCLUSION AND FUTURE DIRECTIONS

Although the potential benefits of AMF symbiosis are well understood, limited studies have been on AMF symbiosis beneath the agricultural system and its applications in crop production practices. These soil microorganisms have the capability to enhance several biophysical and biochemical activities that are associated with soil nutritional constituent, their absorption mechanism, and transportation through various transporters, which leads to nutritional enhancement for sustainable agricultural practices in rice and other cereal crops. This chapter emphasizes aspects of soil quality and nutritional prospects that are influenced by AMF diversity.

REFERENCES

Augé, R.M., Toler, H.D., and Saxton, A.M. (2015). Arbuscular mycorrhizal symbiosis alterss to metal conductance of host plants more under drought than under amply watered conditions: a meta-analysis. *Mycorrhiza* 25, 13–24. doi: 10.1007/s00572-014-0585-4.

Bender, S.F., Plantenga, F., Neftel, A., Jocher, M., Oberholzer, H.R., Köhl, L., et al. (2014). Symbiotic relationships between soil fungi and plants reduce N_2O emissions from soil. *ISMEJ* 8, 1336–1345. doi: 10.1038/ismej.20 13.224.

Breuillin-Sessoms, F., Floss, D.S., Gomez, S.K., Pumplin, N., Ding, Y., Levesque-Tremblay, V., Noar, R.D., Daniels, D.A., Bravo, A., Eaglesham, J.B., and Benedito, V.A. (2015). Suppression of arbuscule degeneration in Medicago truncatula phosphate transporter4 mutants is dependent on the ammonium transporter 2 family protein AMT2; 3. *The Plant Cell*, 27(4), 1352–1366.

Dash, A., and Mohapatra, S.S. (2018). Toxic effect of urea on earthworms determined by simple paper contact method. *Innovare Journal of Agricultural Sciences* 6(1), 1–3.

Delaux, P.M. (2017). Comparative phylogenomics of symbiotic associations. *New Phytologist* 213(1), 89–94.

Gosling, P., Hodge, A., Goodlass, G., and Bending, G.D. (2006). Arbuscular mycorrhizal fungi and organic farming. *Agriculture, Ecosystems & Environment* 113, 17–35. doi: 10.1016/j.agee.2005.09.009.

Igiri, B.E., Okoduwa, S.I.R., Idoko, G.O., Akabuogu, E.P., Adeyi, A.O., and Ejiogu, I.K. (2018). Toxicity and bioremediation of heavy metals contaminated ecosystem from tannery wastewater: a review. *Journal of Toxicology* 2018, 2568038.

Jena, M., Mohapatra, S., and Dash, A. (2018). Yellowness is a threat to newborn: a review. *Asian Journal of Pharmaceutical and Clinical Research* 11, 43. doi: 10.22159/ajpcr.2018.v11i2.22694.

Kobae, Y., Tamura, Y., Takai, S., Banba, M., and Hata, S. (2010). Localized expression of arbuscular mycorrhiza-inducible ammonium transporters in soybean. *Plant and Cell Physiology* 51, 1411–1415. doi: 10.1093/pcp/pcq099.

Ma, J.F., Yamaji, N., Mitani, N., Xu, X.Y., Su, Y.H., McGrath, S.P., and Zhao, F.J. (2008). Transporters of arsenite in rice and their role in arsenic accumulation in rice grain. *Proceedings of the National Academy of Sciences of the United States of America* 105(29):9931–9935. doi: 10.1073/pnas.0802361105.

Meier, S., Cornejo, P., Cartes, P., Borie, F., Medina, J., and Azcón, R. (2015). The interactive effect between Cu-adapted arbuscular mycorrhizal fungi and biotreated agro waste residue to improve the nutritional status of Oenothera picens is growing

Cu-polluted soils. *Journal of Plant Nutrition and Soil Science* 178, 126–135. doi: 10.1002/jpln.201400092.

Mitra, D., Be, G.S., Khoshru, B., De Los Santos Villalobos, S., Belz, C., Chaudhary, P., Shahri, F.N., Djebaili, R., Adeyemi, N.O., El-Ballat, E.M., and El-Esawi, M.A. (2021a). Impacts of arbuscular mycorrhizal fungi on rice growth, development, and stress management with a particular emphasis on strigolactone effects on root development. *Communications in Soil Science and Plant Analysis* 52(14), 1591–1621.

Mitra, D., Djebaili, R., Pellegrini, M., Mahakur, B., Sarker, A., Chaudhary, P., Khoshru, B., Gallo, M.D., Kitouni, M., Barik, D.P., and Panneerselvam, P. (2021b). Arbuscular mycorrhizal symbiosis: plant growth improvement and induction of resistance under stressful conditions. *Journal of Plant Nutrition* 44(13), 1993–2028.

Mitra, D., Rad, K.V., Chaudhary, P., Ruparelia, J., Sagarika, M.S., Boutaj, H., Mohapatra, P.K.D., and Panneerselvam, P. (2021c). Involvement of strigolactone hormone in root development, influence and interaction with mycorrhizal fungi in plant: mini-review. *Current Research in Microbial Sciences* 2, 100026.

Panneerselvam, P., and Saritha, B. (2017). Influence of AM fungi and its associated bacteria on growth promotion and nutrient acquisition in grafted sapota seedling production. *Journal of Applied and Natural Science*, 9(1), 621–625. https://doi.org/10.31018/jans.v9i1.1241

Panneerselvam, P., Kumar, U., Sugitha, T.C.K., Parameswaran, C., Sahoo, S., Binodh, A.K., Jahan, A., and Anandan, A. (2017). Arbuscular mycorrhizal fungi (AMF) for sustainable rice production. In: T. K. Adhya, B. Lal, B. Mohapatra, D. Paul, and S. Das (Eds), *Advances in Soil Microbiology: Recent Trends and Future Prospects* (pp. 99–126). Springer: Singapore.

Panneerselvam, P., Sahoo, S., Senapati, A., Kumar, U., Mitra, D., Parameswaran, C., Anandan, A., Kumar, A., Jahan, A., and Nayak, A.K. (2019). Understanding interaction effect of arbuscular mycorrhizal fungi in rice under elevated carbon dioxide conditions. *Journal of Basic Microbiology* 59(12), 1217–1228.

Paszkowski, U., Kroken, S., Roux, C., and Briggs, S.P., 2002. Rice phosphate transporters include an evolutionarily divergent gene specifically activated in arbuscular mycorrhizal symbiosis. *Proceedings of the National Academy of Sciences*, 99(20), 13324–13329.

Poovarasan, S., Mohandas, S., Paneerselvam, P., Saritha, B., and Ajay, K.M. (2013). Mycorrhizae colonizing actinomycetes promote plant growth and control bacterial blight disease of pomegranate (*Punica granatum* L. cv Bhagwa). *Crop Protection* 53, 175–181.

Pozo, M.J., and Azcón-Aguilar, C. (2007). Unraveling mycorrhiza-induced resistance. *Current Opinion in Plant Biology* 10, 393–398. doi: 10.1016/j.pbi.2007.05.004.

Prasad, R. et al. (2005). Sebacinaceae: culturable mycorrhiza-like endosymbiotic fungi and their interaction with non-transformed and transformed roots. In: S. Declerck, J.A. Fortin, and D.G. Strullu (Eds.), *In Vitro Culture of Mycorrhizas. Soil Biology* (vol. 4). Springer: Berlin, Heidelberg. doi: 10.1007/3-540-27331-X_16.

Rahman, M.A., Parvin, M., Das, U., Ela, E.J., Lee, S.H., Lee, K.W., and Kabir, A.H. (2020). Arbuscular Mycorrhizal symbiosis mitigates iron (Fe)-deficiency retardation in Alfalfa (*Medicago sativa* L.) through the enhancement of Fe accumulation and sulfur-assisted antioxidant defense. *International Journal of Molecular Sciences* 21(6), 2219. doi: 10.3390/ijms21062219.

Ryan, M.H., and Graham, J.H. (2002). Is there a role for arbuscular mycorrhizal fungi in production agriculture? *Plant Soil* 244, 263–271.

Sahoo, S., Panneerselvam, P., Chowdhury, T., Kumar, A., Kumar, U., Jahan, A., Senapati, A., and Anandan, A. (2017). Understanding the AM fungal association in flooded rice under elevated CO_2 condition. *ORYZA-An International Journal on Rice* 54(3), 290–297.

Smith, S.E., and Read, D.J. (2008). *Mycorrhizal Symbiosis*, 3rd Edition. Academic Press: London.

Smith, S.E., and Smith, F.A. (2012). Fresh perspective so the roles of arbuscular mycorrhizal fungi in plant nutrition and growth. *Mycologia* 104, 1–13. doi: 10.3852/11-229.

Smith, S.E., Facelli, E., Pope, S., and Andrew Smith, F. (2009). Plant performance in stressful environments: interpreting new and established knowledge of the roles of arbuscular mycorrhizas. *Plant Soil* 326, 3–20.

Yang, H., Zang, Y., Yuan, Y., Tang, J., and Chen, X., 2012. Selectivity by host plants affects the distribution of arbuscular mycorrhizal fungi: evidence from ITS rDNA sequence metadata. *BMC Evolutionary Biology*, 12, 1–13.

4 Arbuscular Mycorrhizal Fungi

A Sustainable Approach for Enhancing Phosphorous and Nitrogen Use Efficiency in Rice Cultivation

Wiem Alloun
University of Constantine 1

Debasis Mitra
Raiganj University

CONTENTS

4.1 INTRODUCTION

Mycorrhizal symbioses are one of the most promising methods for creating resource-efficient and sustainable agricultural systems. Arbuscular mycorrhizal fungi (AMF) can effectively promote nutrient uptake from the soil, such as

DOI: 10.1201/9781003354086-4

phosphate and nitrogen in many crops, including rice (Panneerselvam et al., 2019; Sahoo et al., 2017).

At the laboratory scale, there is some evidence that the inoculation of limited species of AMF has a positive effect on increasing the concentration of several macro- and micronutrients, which increases photosynthetic activity and, consequently, biomass accumulation (Mitra et al., 2019). Nevertheless, in field conditions, the roots of crops are co-colonized in the field with various AMF species (Panneerselvam et al., 2017), which are challenging to distinguish and can have varying capacities for mineral uptake (Kobae, 2019).

According to several studies, AMF influence the intake of these essential nutrients, primarily N and P, which reduces the uptake of Na and Cl while promoting host growth. These symbiotic fungi actively participate in the cycling of N and P, transferring them to the plants (Mitra et al., 2021). The growth of extraradical mycelium (ERM) makes the uptake possible, directly absorbing amino acids, nitrate, and ammonium. This mechanism increases the efficiency with which plants utilize nitrogen (N) and, as a result, lessens the adverse effects of excessive fertilizer inputs (Basu, Rabara, and Negi, 2018).

In addition, the dynamism of AMF colonization processes and the primarily unknown evolutionary processes make it hard to assess which AMF are functional (Panneerselvam et al., 2020). This chapter discusses significant findings that demonstrate the significance of genomic structure and the dynamics of AM colonization, which may impact AM phosphate absorption properties and P use efficiency (PUE). Additionally, this chapter intends to specify the line of inquiry required to comprehend the function of AMF in the phosphate and nitrogen uptake systems of field crops.

4.2 AGRICULTURE INPUTS AFFECT THE MYCORRHIZAL PHOSPHATE UPTAKE PATHWAY

Environmental factors and the proportion of phosphorus in the soil impact AMF efficiency. There has been evidence of a reduction or inhibition of AMF root colonization and suppression of arbuscule production at high-input phosphate concentrations (Kobae et al., 2016). Additionally, defense mechanisms might help prevent AMF colonization in plants grown under phosphorus-sufficient conditions (Lehnert et al., 2017).

The mycorrhizal association may increase the bioavailability of phosphate in the colonized roots of hosts. According to other investigations, the high AMF colonization improved N and P assimilation, which in turn increased the photosynthetic activity in plant shoots. Additionally, it has been observed that mycorrhizal associations support P and N absorption rates in various irrigation regimes and P-level conditions (Liu et al., 2018).

Cutting-edge omics tools, including functional analysis using RNAseq, NGS, QTL mapping, and docking, are utilized to learn more about the structure, functionality, and physiology of the plant-colonizing AMF population.

4.3 PATTERNS BEHIND AMF CONTRIBUTION TO PLANT'S PHOSPHATE UPTAKE

The AMF's penetration characterizes the plant-AMF symbiosis into cortical root cells, which results in the formation of an arbuscule, densely branching hyphal structure. AMF are placed outside plant cells and surrounded by periarbuscular membranes that connect to their plasma membranes (Harrison and Ivanov, 2017).

These arbuscules actively assist in the exchange of minerals and nutrients between the soil and the roots. In exchange, the fungus receives crucial lipidic and carbohydrate components from the plant host. AMF-plant symbiosis ultimately gives host plants a lot of components (Keymer and Gutjahr, 2018; Lanfranco, Fiorilli, and Gutjahr, 2018).

According to several investigations, AMF significantly increases a plant's ability to absorb inorganic nutrients from nutrient-poor soils, particularly phosphate (Kayama and Yamanaka, 2014). Greater leaf area and nitrogen, potassium, calcium, and phosphorus contents were observed in experimental trials on tomato plants inoculated with AMF, demonstrating accelerated plant growth brought on by the host roots' increased surface-absorbing capacity of host roots (Balliu, Sallaku, and Rewald, 2015).

Nevertheless, based on the species of AMF, various mycorrhizal phosphate uptake levels may be determined. Due to this, compared with uninoculated plants, the phosphate uptake in inoculated plants does not increase. The level of mycorrhizal phosphate uptake varies among the different forms of AMF, which may be the cause of this level disparity. The performance of phosphate uptake across AMF species differs noticeably, according to a study on the inoculation of *Medicago sativa* with more than 30 different types (Mensah et al., 2015).

AMF increase phosphate uptake from arbuscules by inducing the plant host genes for phosphate transporters, and the produced protein is located on the periarbuscular membrane (Pumplin et al., 2012). Premature degradation of the arbuscules and a decrease in total phosphate uptake have been noted in the symbiotic phosphate transporter genes mutant plants. These changes supported the upregulation of the symbiotic phosphate transport system, indicating the crucial role of the phosphate transport system in establishing the phosphate uptake *mycorrhizal pathway* in plant roots (Willmann et al., 2013).

Additionally, it was proposed that the epidermal "direct pathway" of phosphate uptake involves the downregulation of phosphate transporter genes (Tamura et al., 2012). However, more investigations are required to determine the mechanisms behind the balanced contribution of both pathways.

4.4 AMF-MEDIATED NITROGEN TRANSLOCATION INTO PLANTS

Several studies have advocated the role of AMF in absorbing and enhancing the acquisition of mineral nutrients, such as N, in host plants, thus improving their growth (Turrini et al., 2018). The AMF assemblages constituted of hyphal networks limit the inputs of N in soils by altering the dynamics of microbial activity in their host's rhizospheric. This activity includes regulating the N cycling contributors, denitrifying, nitrifying, and free-living bacteria (Veresoglou, Chen, and Rillig, 2012).

Compared to non-inoculated plants, AMF have shown that the increase in N accumulation in plant tissues is responsible for the rise in dry biomass output (Zhu et al., 2016). For instance, in conditions such as low N and P levels, the AMF extraradical hyphae extending from roots to the environment absorb and distribute inorganic N and P into the plant, which increases shoot biomass, panicles, and grain production (Zhang et al., 2018). Additionally, the transfer of N to the plant by symbiotic AMF increases photosynthetic activity, which is reflected in a more significant amount of chlorophyll in their leaves, given that chlorophyll molecules are adept at trapping N (de Andrade et al., 2015).

AMF as well as plants behave differently in the presence of external *stimuli*. It has been speculated that AMF symbiosis improves C and N accumulation and assimilation under elevated CO_2 concentrations (Zhu et al., 2016; Panneerselvam et al., 2019). Furthermore, salt stress disturbs the P and N levels individually and the N:P ratio in plant shoots (Wang et al., 2018). Consequently, AMF increases N levels in their host plants during exposure to salinity (Turrini et al., 2018).

4.5 AMF IMPLICATIONS IN NITROGEN USE EFFICIENCY IN PLANTS

AMF play a pivotal role in the plant's mineral nutrition, including rice, and their symbiosis is mainly based on the complex chemical signaling pathways between both parties. AMF communicate with plants through strigolactones (SL). These phytohormones regulate root development and shoot branching and constitute signal molecules in modulating rice plant root growth and assisting fungal symbiotic relationships (Mitra et al., 2021). As a consequence, AMF improve the bioavailability, distribution, and translocation of these essential nutrients, which are the limiting factor, boost the P and N acquisition, and promote their growth (Luo et al., 2018; Tanaka et al., 2022). AMF also improve the nutrient use efficiency of the host (Xing et al., 2019). The benefit behind this AMF-mediated uptake facilitation of N and P resides in the vast array of root exudates of their host plants (Basu, Rabara, and Negi, 2018).

Field meta-analysis suggested the positive effect of AMF interconnected networks in improving N uptake and NUE from and in plants. These findings advocated that the employment of AMF as plant inoculants may be beneficial for crops and help in increasing the nitrogen use efficiency (NUE) in plants (Mensah et al., 2015).

4.6 RICE-AMF ASSOCIATION AFFECTS N AND P UPTAKE

Glomeromycota is the group of dominating symbionts that colonize rice roots. These symbiotic fungi generate extensive extraradical hyphae and supply sufficient amounts of vital N and P to rice (Bao, Wang, and Olsson, 2019). Thus,

AMF minimize the detrimental effects of excessive agricultural inputs and take advantage of plant root exudates necessary for optimum growth. The close association between AMF and the rice plants they live on increases biomass output and reduces biotic and abiotic stressors (Campo et al., 2020; Tisarum et al., 2020).

Additionally, AMF interact with other microbes in the rice plant's mycorrhizosphere. This three-way symbiosis is essential for the health of rice plants, microbial diversity, and soil properties. However, several microbial pathways, root exudates, microbial metabolites, and chemical signals are necessary for this symbiosis and rice root colonization.

Nevertheless, the nature and function of this association need to be better understood. Future research goals may be determined by the pathways implicated in rice-AMF interaction and mutualisms and their actual influence on rice plant performance.

4.7 DIRECTING THE NEXT RESEARCH PANEL TO AMF-RICE SIGNALING PATHWAYS

The evidence about the positive effect of AMF in nutrient uptake has been intensively studied; however, its role in nutrient signaling, ion transport, and nutrient channels in rice plants is not explored. Thus, the role of AMF in translocation and enhancement of the NUE mechanism is poorly studied; therefore, it is important to determine how AMF affect the movement of N and P in field conditions. Hence, it would be interesting if research focuses more on the signaling pathways involved in the communication between rice and their AM symbionts, and the molecular factors implicated in cycling mineral nutrients and regulating their levels and consequently affecting the photosynthetic activity of the host plant.

Similarly, a substantial research area may be found in identifying host plant and AMF-specific protein factors that regulate the symbiotic interaction and essential cellular and metabolic pathways. In plants, the influence of AMF on nitrogen use efficiency (NUE) has been thoroughly studied.

Extensive research at the transcriptomic, proteomic, and metabolomic levels must clarify the function of AMF in N and P uptake and its impact on plant metabolism. More research should focus on the mechanisms underlying the interactions between AMF and other soil microorganisms and the interactive activities between AMF and soil microorganisms in rice.

4.8 CONCLUSION

AMF significantly enhance the growth and development of plants. The abundance of compounds exuded from rice roots, which also support a variety of symbiotic associations in the rice mycorrhizosphere, favor AMF inoculation and establishment in rice plants. The involvement of these symbionts in soil nutrient cycling, which enhances nutrient uptake and nutrient use efficiency (PUE and NUE) and reduces the adverse effects of significant agricultural inputs, leads to an AMF-mediated enhancement in plant growth and production.

REFERENCES

Balliu, Astrit, Glenda Sallaku, and Boris Rewald. 2015. "AMF Inoculation Enhances Growth and Improves the Nutrient Uptake Rates of Transplanted, Salt-Stressed Tomato Seedlings." *Sustainability (Switzerland)* 7(12): 15967–81. https://doi.org/10.3390/su71215799.

Bao, Xiaozhe, Yutao Wang, and Pål Axel Olsson. 2019. "Arbuscular Mycorrhiza under Water: Carbon–phosphorus Exchange between Rice and Arbuscular Mycorrhizal Fungi under Different Flooding Regimes." *Soil Biology and Biochemistry* 129: 169–77. https://doi.org/10.1016/j.soilbio.2018.11.020.

Basu, Supratim, Roel C. Rabara, and Sangeeta Negi. 2018. "AMF: The Future Prospect for Sustainable Agriculture." *Physiological and Molecular Plant Pathology*. https://doi.org/10.1016/j.pmpp.2017.11.007.

Campo, Sonia, Héctor Martín-Cardoso, Marta Olivé, Eva Pla, Mar Catala-Forner, Maite Martínez-Eixarch, and Blanca San Segundo. 2020. "Effect of Root Colonization by Arbuscular Mycorrhizal Fungi on Growth, Productivity and Blast Resistance in Rice." *Rice* 13(1). https://doi.org/10.1186/s12284-020-00402-7.

de Andrade, Sara Adrian Lopez, Adilson Pereira Domingues, and Paulo Mazzafera. 2015. "Photosynthesis Is Induced in Rice Plants That Associate with Arbuscular Mycorrhizal Fungi and Are Grown under Arsenate and Arsenite Stress." *Chemosphere* 134: 141–49. https://doi.org/10.1016/j.chemosphere.2015.04.023.

Harrison, Maria J., and Sergey Ivanov. 2017. "Exocytosis for Endosymbiosis: Membrane Trafficking Pathways for Development of Symbiotic Membrane Compartments." *Current Opinion in Plant Biology*. https://doi.org/10.1016/j.pbi.2017.04.019.

Kayama, Masazumi, and Takashi Yamanaka. 2014. "Growth Characteristics of Ectomycorrhizal Seedlings of Quercus Glauca, Quercus Salicina, and Castanopsis Cuspidata Planted on Acidic Soil." *Trees - Structure and Function* 28(2): 569–83. https://doi.org/10.1007/s00468-013-0973-y.

Keymer, Andreas, and Caroline Gutjahr. 2018. "Cross-Kingdom Lipid Transfer in Arbuscular Mycorrhiza Symbiosis and Beyond." *Current Opinion in Plant Biology*. https://doi.org/10.1016/j.pbi.2018.04.005.

Kobae, Yoshihiro. 2019. "Dynamic Phosphate Uptake in Arbuscular Mycorrhizal Roots under Field Conditions." *Frontiers in Environmental Science*. Frontiers Media S.A. https://doi.org/10.3389/fenvs.2018.00159.

Kobae, Yoshihiro, Yoshihiro Ohmori, Chieko Saito, Koji Yano, Ryo Ohtomo, and Toru Fujiwara. 2016. "Phosphate Treatment Strongly Inhibits New Arbuscule Development but Not the Maintenance of Arbuscule in Mycorrhizal Rice Roots." *Plant Physiology* 171(1): 566–79. https://doi.org/10.1104/pp.16.00127.

Lanfranco, Luisa, Valentina Fiorilli, and Caroline Gutjahr. 2018. "Partner Communication and Role of Nutrients in the Arbuscular Mycorrhizal Symbiosis." *New Phytologist*. Blackwell Publishing Ltd. https://doi.org/10.1111/nph.15230.

Lehnert, Heike, Albrecht Serfling, Matthias Enders, Wolfgang Friedt, and Frank Ordon. 2017. "Genetics of Mycorrhizal Symbiosis in Winter Wheat (Triticum Aestivum)." *New Phytologist* 215(2): 779–91. https://doi.org/10.1111/nph.14595.

Liu, Caixia, Sabine Ravnskov, Fulai Liu, Gitte H. Rubæk, and Mathias N. Andersen. 2018. "Arbuscular Mycorrhizal Fungi Alleviate Abiotic Stresses in Potato Plants Caused by Low Phosphorus and Deficit Irrigation/Partial Root-Zone Drying." *Journal of Agricultural Science* 156(1): 46–58. https://doi.org/10.1017/S0021859618000023.

Luo, Le, Hongxuan Wang, Xiaohong Liu, Jinqi Hu, Xueli Zhu, Shou Pan, Ruyi Qin, et al. 2018. "Strigolactones Affect the Translocation of Nitrogen in Rice." *Plant Science* 270: 190–7. https://doi.org/10.1016/j.plantsci.2018.02.020.

Mensah, Jerry A., Alexander M. Koch, Pedro M. Antunes, Toby Kiers, Miranda Hart, and Heike Bücking. 2015. "High Functional Diversity within Species of Arbuscular Mycorrhizal Fungi Is Associated with Differences in Phosphate and Nitrogen Uptake and Fungal Phosphate Metabolism." *Mycorrhiza* 25(7): 533–46. https://doi.org/10.1007/s00572-015-0631-x.

Mitra, Debasis, Beatriz Elena Guerra Sierra, Bahman Khoshru, Sergio De Los Santos Villalobos, Claudia Belz, Priya Chaudhary, Faride Noroozi Shahri, et al. 2021. "Impacts of Arbuscular Mycorrhizal Fungi on Rice Growth, Development, and Stress Management with a Particular Emphasis on Strigolactone Effects on Root Development." *Communications in Soil Science and Plant Analysis* 52(14): 1591–621. https://doi.org/10.1080/00103624.2021.1892728.

Mitra, Debasis, Navendra Uniyal, Periyasamy Panneerselvam, Ansuman Senapati, Arakalagud Nanjundaiah Ganeshamurthy, and Divya Jain. 2019. "Role of Mycorrhiza and Its Associated Bacteria on Plant Growth Promotion and Nutrient Management in Sustainable Agriculture." *IJLSAS* 1(1): 1–10. http://www.ijlsas.com/?mno=302644881.

Panneerselvam, Periyasamy, Sowarnalisha Sahoo, Ansuman Senapati, Upendra Kumar, Debasis Mitra, Chidambaranathan Parameswaran, Annamalai Anandan, Anjani Kumar, Afrin Jahan, and Amaresh Kumar Nayak. 2019. "Understanding Interaction Effect of Arbuscular Mycorrhizal Fungi in Rice under Elevated Carbon Dioxide Conditions." *Journal of Basic Microbiology* 59(12): 1217–28.

Panneerselvam, Periyasamy, Upendra Kumar, Ansuman Senapati, Chidambaranathan Parameswaran, Annamalai Anandan, Anjani Kumar, Afrin Jahan, S. R. Padhy, and Amaresh Kumar Nayak. 2020. "Influence of Elevated CO_2 on Arbuscular Mycorrhizal Fungal Community Elucidated Using Illumina MiSeq Platform in Sub-Humid Tropical Paddy Soil." *Applied Soil Ecology* 145: 103344.

Panneerselvam, Periyasamy, Upendra Kumar, Sugitha Thankappan, Chidambaranathan Parameswaran, Sowarnalisha Sahoo, Asish Kanakaraj Binodh, Afrin Jahan, and Annamalai Anandan. 2017. Arbuscular mycorrhizal fungi (AMF) for sustainable rice production. In: Tapan Kumar Adhya, Banwari Lal, Balaram Mohapatra, Dhiraj Paul, Subhasis Das (Eds.), *Advances in Soil Microbiology: Recent Trends and Future Prospects* (pp. 99–126). Springer, Singapore.

Pumplin, Nathan, Xinchun Zhang, Roslyn D. Noar, and Maria J. Harrison. 2012. "Polar Localization of a Symbiosis-Specific Phosphate Transporter Is Mediated by a Transient Reorientation of Secretion." *Proceedings of the National Academy of Sciences of the United States of America* 109(11). https://doi.org/10.1073/pnas.1110215109.

Sahoo, Sowarnalisha, Periyasamy Panneerselvam, Tapas Chowdhury, Anjani Kumar, Upendra Kumar, Afrin Jahan, Ansuman Senapati, and Annamalai Anandan. 2017. "Understanding the AM Fungal Association in Flooded Rice under Elevated CO_2 Condition." *ORYZA-An International Journal on Rice* 54(3): 290–7.

Tamura, Yosuke, Yoshihiro Kobae, Toyotaka Mizuno, and Shingo Hata. 2012. "Identification and Expression Analysis of Arbuscular Mycorrhiza-Inducible Phosphate Transporter Genes of Soybean." *Bioscience, Biotechnology and Biochemistry* 76(2): 309–13. https://doi.org/10.1271/bbb.110684.

Tanaka, Sachiko, Kayo Hashimoto, Yuuki Kobayashi, Koji Yano, Taro Maeda, Hiromu Kameoka, Tatsuhiro Ezawa, Katsuharu Saito, Kohki Akiyama, and Masayoshi Kawaguchi. 2022. "Asymbiotic Mass Production of the Arbuscular Mycorrhizal Fungus Rhizophagus Clarus." *Communications Biology* 5(1). https://doi.org/10.1038/s42003-021-02967-5.

Tisarum, Rujira, Cattarin Theerawitaya, Thapanee Samphumphuang, Kanyamin Polispitak, Panarat Thongpoem, Harminder Pal Singh, and Suriyan Cha-um. 2020. "Alleviation of Salt Stress in Upland Rice (*Oryza Sativa* L. Ssp. Indica Cv. Leum Pua) Using Arbuscular Mycorrhizal Fungi Inoculation." *Frontiers in Plant Science* 11. https://doi.org/10.3389/fpls.2020.00348.

Turrini, Alessandra, Alberico Bedini, Mario Bonilla Loor, Gaia Santini, Cristiana Sbrana, Manuela Giovannetti, and Luciano Avio. 2018. "Local Diversity of Native Arbuscular Mycorrhizal Symbionts Differentially Affects Growth and Nutrition of Three Crop Plant Species." *Biology and Fertility of Soils* 54(2): 203–17. https://doi.org/10.1007/s00374-017-1254-5.

Veresoglou, Stavros D., Baodong Chen, and Matthias C. Rillig. 2012. "Arbuscular Mycorrhiza and Soil Nitrogen Cycling." *Soil Biology and Biochemistry*. https://doi.org/10.1016/j.soilbio.2011.11.018.

Wang, Yanhong, Minqiang Wang, Yan Li, Aiping Wu, and Juying Huang. 2018. "Effects of Arbuscular Mycorrhizal Fungi on Growth and Nitrogen Uptake of Chrysanthemum Morifolium under Salt Stress." *PLoS One* 13(4). https://doi.org/10.1371/journal.pone.0196408.

Willmann, Martin, Nina Gerlach, Benjamin Buer, Aleksandra Polatajko, Reka Nagy, Eva Koebke, Jan Jansa, Rene Flisch, and Marcel Bucher. 2013. "Mycorrhizal Phosphate Uptake Pathway in Maize: Vital for Growth and Cob Development on Nutrient Poor Agricultural and Greenhouse Soils." *Frontiers in Plant Science* 4. https://doi.org/10.3389/fpls.2013.00533.

Xing, Yingying, Wenting Jiang, Xiaolong He, Sajid Fiaz, Shakeel Ahmad, Xin Lei, Wenqiang Wang, Yanfeng Wang, and Xiukang Wang. 2019. "A Review of Nitrogen Translocation and Nitrogen-Use Efficiency." *Journal of Plant Nutrition*. Taylor & Francis Inc. https://doi.org/10.1080/01904167.2019.1656247.

Zhang, Fei, Jia Dong He, Qiu Dan Ni, Qiang Sheng Wu, and Ying Ning Zou. 2018. "Enhancement of Drought Tolerance in Trifoliate Orange by Mycorrhiza: Changes in Root Sucrose and Proline Metabolisms." *Notulae Botanicae Horti Agrobotanici Cluj-Napoca* 46(1): 270–6. https://doi.org/10.15835/nbha46110983.

Zhu, Xiancan, Fengbin Song, Shengqun Liu, and Fulai Liu. 2016. "Arbuscular Mycorrhiza Improve Growth, Nitrogen Uptake, and Nitrogen Use Efficiency in Wheat Grown under Elevated CO_2." *Mycorrhiza* 26(2): 133–40. https://doi.org/10.1007/s00572-015-0654-3.

5 Arbuscular Mycorrhizal Fungi and Their Role in Plant Growth Promotion in Rice

Ankita Priyadarshini, Suchismita Behera, Debasis Mitra and Ansuman Senapati
ICAR - National Rice Research Institute

Swagat Shubhadarshi
Siksha 'O' Anusandhan University

Sucharita Satapathy
ICAR - National Rice Research Institute

Subhadra Pattanayak
Regional Research and Technology Transfer Stations

Periyasamy Panneerselvam
ICAR - National Rice Research Institute

CONTENTS

5.1 INTRODUCTION

Mycorrhiza comprises two words: *myco* refers to fungi, and *rhiza* refers to a plant's root system. Approximately 70%–90% of land plant species have mutualistic

DOI: 10.1201/9781003354086-5

symbiotic relationships with fungi that are members of the monophyletic phylum Glomeromycota, known as arbuscular mycorrhizal fungi (AMF). Most AMF species are members of the phylum Mucoromycota's subphylum Glomeromycotina (Begum et al., 2019). This subphylum contains 25 genera and four orders of AMF, namely, Glomerales, Archaeosporales, Paraglomerales, and Diversisporales (Redecker et al., 2013). AMF are soilborne fungi that reside in the soil and significantly increase plant nutrient uptake and resistance to several abiotic stresses (Sun et al., 2018). In exchange for photosynthetically fixed carbon, root colonization by AMF enhances the host plant's ability to absorb mineral nutrients, particularly phosphorus and nitrogen; reduces the toxicity of heavy metals in the host plants; and enables the host plants to withstand high metal concentrations in the soil. Moreover, it has been indicated that root colonization by AMF enhances nutrition, which ultimately aids in plant growth and development. Improved resistance to biotic and abiotic stress is a benefit of AM symbiosis in several plant species (Wang et al., 2018).

Most rice is produced worldwide in wetland ecosystems that are anaerobic and submerged. Therefore, field management that promotes the growth of AMF symbioses and probably even more AMF inoculation could enhance AMF colonization in rice (Vallino et al., 2009). It has been observed that the use of AMF at the nursery stage increases rice yield by 14%–21% (Solaiman and Hirata, 1997). The interaction between rice plants and AMF will assist in the development of rice-based AMF bio-inoculants for the effective management of P and other nutrients under various cultivation conditions (Panneerselvam et al., 2017). AMF enhance photosynthesis by maintaining higher levels of chlorophyll fluorescence and reducing stomatal closure, especially during drought conditions. AMF, therefore, encourage better drought recovery, which results in higher rice grain yields. Furthermore, AMF act as a biocontrol agent to minimize the necessity of pesticides, increase root water uptake, provide enough water to maintain plant physiological activity, and change the toxicity brought on by salt stress. They encourage plant growth and development in an environmentally sustainable manner.

5.2 ROLE IN PLANT GROWTH PROMOTION IN RICE

Rice (*Oryza sativa* L.) is a staple food that provides nutrition to more than half of the world's population. Its demand is rising as the world's population expands. Recently, the importance of sustainable agriculture has been widely recognized, and it has emerged as the most crucial issue in agriculture. Plant nutrition in low-fertilized sustainable agrosystems can thus be improved using AM. AMF can also act as a biocontrol agent to reduce pesticide use, improve root water uptake, provide adequate water to preserve plant physiological activity, and alter the toxicity induced by salt stress, which promotes plant growth and development in an eco-friendly manner.

5.2.1 The Role of AMF in Nutrient Management

Rice is usually grown in several ecosystems; 78% of the world's rice is grown in lowland regions under irrigated and rainfed conditions. In comparison to other crop species, rice fields use more fertilizers. AMF create symbiotic relationships with host plant roots to get essential nutrients and, in exchange, provide mineral nutrients, both macro- and micronutrients such as N, P, K, Ca, Zn, and S. As a result, AMF provide nutritional support to plants even when conditions inside the root cells are unfavorable. Several investigations have shown that AMF can absorb and transfer N to plants or host plants (Turrini et al., 2018). According to plant nutritional requirements, AMF contribute significantly more plant phosphorus than plant nitrogen absorption. Phosphorus (P), a vital nutrient, participates in numerous cellular and physiological mechanisms in a living body. The rhizosphere modification solubilizes soil phosphorus by releasing phosphatase enzymes, organic acids, and metabolites such as siderophores, inhibiting P nutrition uptake and acquisition (Shenoy and Kalagudi, 2005). Root colonization by the AM fungus increases plant nutrition by improving the availability and transfer of nutrients, particularly phosphate (Pi), to host plant roots via the mycelium and hyphae (Campo et al., 2020). The host plant-AMF association alters the properties of the rhizosphere by changing the pH and root exudation profile, enhancing phosphate solubilization for developing the microbial community (Mitra et al., 2021). In addition to macronutrients, the AMF association increased the phytoavailability of micronutrients like zinc and copper (Begum et al., 2019).

Mycorrhizal inoculum-treated rice with an aerobic genotype indicated 28%–57% greater root colonization (Gao et al., 2007). AMF colonization helped upland rice seedlings to produce higher yields at a suitable fertilizer dosage, reducing costs and environmental damage in the process. Increased N and P redistribution to panicles by AMF increased the allocation of shoot biomass to panicles and grains, especially when fertilizer levels were low. From heading to maturity, there is an increased N translocation into seeds (Zhang et al., 2017). As a result, AMF, which are naturally occurring root symbionts, give host plants vital inorganic plant nutrients, which enhance growth and yield in both unstressed and stressed conditions.

5.2.2 The Role of AMF as a Biocontrol Agent

AMF are used as a natural biocontrol agent because they enhance antagonistic interactions with different kinds of soilborne plant pathogens, including *Phytophthora* sp., *Rhizoctonia solani* (Rice-sheath blight), and *Pythium ultimum* (Mitra et al., 2021). AMF can influence the physical and chemical properties of soil, promote the growth of other beneficial microorganisms, and compete with pathogenic microorganisms to improve the rhizosphere environment (Weng et al., 2022). AMF are a promising substitute because they compete with pathogenic plant organisms for nutrients and space (Berg et al., 2007). Through their potential as a biocontrol agent, AMF increase competition for the host and colonization site; changes in the root system's anatomy; plant nutrient absorption; biochemical and physiochemical alterations; and activation of defense mechanisms in plants (Mitra et al., 2021).

5.2.3 The Role of AMF in Rice Drought Stress Management

Rice is grown in four different ecosystems worldwide, which are limited by irrigation water availability. Water scarcity, on the contrary, has a negative impact on rice production (Singh et al., 2021). Increased plant water acquisition can be achieved by inoculating AMF symbiotically associated with plant roots due to the large volume of soil explored by the roots and the extraradical hyphae of the fungi (Gianinazzi et al., 2010), thus increasing crop productivity and plant growth under stressed conditions (Barea et al., 2005; Azcón and Barea 2010). Rice root colonization by AMF is stimulated by non-flooded rice cultivation and the prevailing aerobic environment in the soil (Panneerselvam et al., 2017, 2019). Inoculation with AMF increased ascorbic and proline levels, which are very effective protective compounds against the negative effects of water scarcity. Under drought conditions, the symbiotic relationship of various plants with AMF may eventually improve root size and efficiency, leaf area index, and biomass (Gholamhoseini et al., 2013). Moreover, the AMF symbiosis improves gas exchange, leaf water relations, stomatal conductance, and transpiration rate (Mena-Violante et al., 2006; Sahoo et al., 2017). As a result, AMF are useful for rice plants as an environmentally friendly and adequate technology for improving plant growth and productivity.

5.3 CONCLUSIONS AND FUTURE PROSPECTS

The AMF symbiotic interaction has been widely accepted for enhancing plant growth in challenging situations and enhancing rice growth and soil structural features substantially. Plants associated with AMF can thrive in the presence of biotic and abiotic stresses such as nutrient depletion, drought, and salinity. This knowledge related to the AMF function was combined cohesively in this study to explain the symbiotic link between AMF and plant diversity under stress in rice and other plant systems. The use of AMF in plant disease control is bound to become a viable and eco-friendly strategy for reducing pathogen occurrence and achieving green and sustainable development. The application of AMF for agricultural enhancement can greatly minimize the use of synthetic fertilizers and other chemicals, encouraging bio-healthy farming. AMF-mediated development and productivity enhancement in crop plants may be effective in satisfying the worldwide consumption needs of an increasing population. The adoption of ecologically sustainable technology, on the contrary, is strongly encouraged. More research should be conducted to discover genes that regulate AMF-mediated growth and development under stressful situations. Modulations in tolerance mechanisms and crosstalk triggered to regulate plant performance caused by AMF can assist in enhancing crop yield. The AMF must be examined at all levels to become a sustainable organic fertilizer in the near future.

REFERENCES

Azcón, R. and Barea, J.M., 2010. Mycorrhizosphere interactions for legume improvement. In M.S. Khan, A. Zaidi and J. Musarrat (Eds.), *Microbes for Legume Improvement* (pp. 237–271). Springer, Vienna.

Barea, J.M., Pozo, M.J., Azcon, R. and Azcon-Aguilar, C., 2005. Microbial co-operation in the rhizosphere. *Journal of Experimental Botany*, *56*(417), pp. 1761–1778.

Begum, N., Qin, C., Ahanger, M.A., Raza, S., Khan, M.I., Ashraf, M., Ahmed, N. and Zhang, L., 2019. Role of arbuscular mycorrhizal fungi in plant growth regulation: implications in abiotic stress tolerance. *Frontiers in Plant Science*, *10*, p. 1068.

Berg, G., Grosch, R., Scherwinski, K., Pujol, M., Badosa, E., Manceau, C., Luchi, N., Capretti, P. and Pinzani, P., 2007. Risk assessment for microbial antagonists: are there effects on. *Gesunde Pflanz*, *59*, pp. 107–117.

Campo, S., Martín-Cardoso, H., Olivé, M., Pla, E., Catala-Forner, M., Martínez-Eixarch, M. and San Segundo, B., 2020. Effect of root colonization by arbuscular mycorrhizal fungi on growth, productivity and blast resistance in rice. *Rice*, *13*(1), pp. 1–14.

Gao, X., Kuyper, T.W., Zou, C., Zhang, F. and Hoffland, E., 2007. Mycorrhizal responsiveness of aerobic rice genotypes is negatively correlated with their zinc uptake when nonmycorrhizal. *Plant and Soil*, *290*(1), pp. 283–291.

Gholamhoseini, M., Ghalavand, A., Dolatabadian, A., Jamshidi, E. and Khodaei-Joghan, A., 2013. Effects of arbuscular mycorrhizal inoculation on growth, yield, nutrient uptake and irrigation water productivity of sunflowers grown under drought stress. *Agricultural Water Management*, *117*, pp. 106–114.

Gianinazzi, S., Gollotte, A., Binet, M.N., van Tuinen, D., Redecker, D. and Wipf, D., 2010. Agroecology: the key role of arbuscular mycorrhizas in ecosystem services. *Mycorrhiza*, *20*(8), pp. 519–530.

Mena-Violante, H.G., Ocampo-Jiménez, O., Dendooven, L., Martínez-Soto, G., González-Castañeda, J., Davies, F.T. and Olalde-Portugal, V., 2006. Arbuscular mycorrhizal fungi enhance fruit growth and quality of chile ancho (*Capsicum annuum L.* cv San Luis) plants exposed to drought. *Mycorrhiza*, *16*(4), pp. 261–267.

Mitra, D., Be, G.S., Khoshru, B., De Los Santos Villalobos, S., Belz, C., Chaudhary, P., Shahri, F.N., Djebaili, R., Adeyemi, N.O., El-Ballat, E.M. and El-Esawi, M.A., 2021. Impacts of arbuscular mycorrhizal fungi on rice growth, development, and stress management with a particular emphasis on strigolactone effects on root development. *Communications in Soil Science and Plant Analysis*, *52*(14), pp. 1591–1621.

Panneerselvam, P., Kumar, U., Sugitha, T.C.K., Parameswaran, C., Sahoo, S., Binodh, A.K., Jahan, A. and Anandan, A., 2017. Arbuscular mycorrhizal fungi (AMF) for sustainable rice production. In T.K. Adhya, B. Lal, B. Mohapatra, D. Paul and S. Das (Eds.), *Advances in Soil Microbiology: Recent Trends and Future Prospects* (pp. 99–126). Springer, Singapore.

Panneerselvam, P., Sahoo, S., Senapati, A., Kumar, U., Mitra, D., Parameswaran, C., Anandan, A., Kumar, A., Jahan, A. and Nayak, A.K., 2019. Understanding interaction effect of arbuscular mycorrhizal fungi in rice under elevated carbon dioxide conditions. *Journal of Basic Microbiology*, *59*(12), pp. 1217–1228.

Redecker, D., Schüßler, A., Stockinger, H., Stürmer, S.L., Morton, J.B. and Walker, C., 2013. An evidence-based consensus for the classification of arbuscular mycorrhizal fungi (Glomeromycota). *Mycorrhiza*, *23*(7), pp. 515–531.

Sahoo, S., Panneerselvam, P., Chowdhury, T., Kumar, A., Kumar, U., Jahan, A., Senapati, A. and Anandan, A., 2017. Understanding the AM fungal association in flooded rice under elevated CO_2 condition. *ORYZA-An International Journal on Rice*, *54*(3), pp. 290–297.

Shenoy, V.V. and Kalagudi, G.M., 2005. Enhancing plant phosphorus use efficiency for sustainable cropping. *Biotechnology Advances*, *23*(7–8), pp. 501–513.

Singh, B., Mishra, S., Bisht, D.S. and Joshi, R., 2021. Growing rice with less water: improving productivity by decreasing water demand. In J. Ali and S.H. Wani (Eds), *Rice Improvement* (pp. 147–170). Springer, Cham.

Solaiman, M.Z. and Hirata, H., 1997. Effect of arbuscular mycorrhizal fungi inoculation of rice seedlings at the nursery stage upon performance in the paddy field and greenhouse. *Plant and Soil*, *191*(1), pp. 1–12.

Sun, Z., Song, J., Xin, X.A., Xie, X. and Zhao, B., 2018. Arbuscular mycorrhizal fungal 14-3-3 proteins are involved in arbuscule formation and responses to abiotic stresses during AM symbiosis. *Frontiers in Microbiology*, *9*, p. 91.

Turrini, A., Bedini, A., Loor, M.B., Santini, G., Sbrana, C., Giovannetti, M. and Avio, L., 2018. Local diversity of native arbuscular mycorrhizal symbionts differentially affects growth and nutrition of three crop plant species. *Biology and Fertility of Soils*, *54*(2), pp. 203–217.

Vallino, M., Greppi, D., Novero, M., Bonfante, P. and Lupotto, E., 2009. Rice root colonisation by mycorrhizal and endophytic fungi in aerobic soil. *Annals of Applied Biology*, *154*(2), pp. 195–204.

Wang, Y., Wang, M., Li, Y., Wu, A. and Huang, J., 2018. Effects of arbuscular mycorrhizal fungi on growth and nitrogen uptake of Chrysanthemum morifolium under salt stress. *PLoS One*, *13*(4), p. e0196408.

Weng, W., Yan, J., Zhou, M., Yao, X., Gao, A., Ma, C., Cheng, J. and Ruan, J., 2022. Roles of Arbuscular mycorrhizal fungi as a biocontrol agent in the control of plant diseases. *Microorganisms*, *10*(7), p. 1266.

Zhang, X., Wang, L., Ma, F., Yang, J. and Su, M., 2017. Effects of arbuscular mycorrhizal fungi inoculation on carbon and nitrogen distribution and grain yield and nutritional quality in rice (*Oryza sativa* L.). *Journal of the Science of Food and Agriculture*, *97*(9), pp. 2919–2925.

6 Arbuscular Mycorrhizal Fungi and Strigolactone

Role, Application, and Effects of Synthetic Strigolactone in Plant Growth Promotion

Partha Chandra Mondal and Shreosi Biswas
Indian Agricultural Research Institute

Puranjoy Sar and Biswajit Pramanik
Palli Siksha Bhavana (Institute of Agriculture)

CONTENTS

6.1 INTRODUCTION

Arbuscular mycorrhizal fungi (AMF), a group of obligate parasitic microbes, associate with more than 80% of terrestrial plants symbiotically (Harrison, 2005). These fungi enter plant roots and colonize them, developing arbuscules, which are highly branched structures and centers of nutrient exchange. In exchange, the AMF acquire photosynthates from their host plants. The fungi provide their hosts with water and nutrients, principally phosphate and nitrogen that are collected through the hyphae found in the soil. The most prevalent and pervasive symbiosis

DOI: 10.1201/9781003354086-6

on earth, AM symbiosis has existed over 460 million years (Simon et al., 1993). It is thought that the development of this synergism was crucial for plants to colonize the land. The plant becomes more resilient to biotic and abiotic stresses because of AM symbiosis as well. The ability of the AMF to quickly attach with plant roots and establish symbiotic interactions with them is essential for their survival as obligate biotrophs. AMF's spores can develop without a host (non-symbiotic growth), but the root exudates play a major role in growing and branching of the germinating hyphae as well as infection of the root (pre-symbiotic growth) (Besserer et al., 2006; Mitra et al., 2021a,b). Along with large amounts of CO_2, inorganic ions, amino acids, organic acids, sugars, vitamins, purines, nucleosides, enzymes, and hormones, roots also secrete small amounts of various secondary metabolites, including flavonoids, terpenoids, and other metabolites, that act as early chemical mediators of more specific plant-microbe interactions. Strigolactones are a group of apocarotenoids that act as signaling compounds in the rhizosphere. In addition to stimulating the germination of parasitic weeds like *Striga* and *Orobanche*, they play a crucial role in spore germination and hyphal branching of AMF, which further induce the symbiosis between AMF and higher plants.

6.2 BRANCHING FACTORS TO STRIGOLACTONES

Among the various chemicals present in root exudates, flavonoids were thought to promote hyphal branching at first. However, this notion was denied when maize mutants deficient in flavonoids showed normal AMF colonization (Becard et al., 1995; Mitra et al., 2021c,d). Previous research revealed that a branching factor (BF) derived from hairy root cultures of *Daucus carota* greatly encourages the branching of *Gigaspora* spp. germinating hyphae, a response akin to that commonly seen when hyphae of AMF develop in the presence of living roots (Nagahashi and Douds, 1999; Buee et al., 2000). The strigolactone 5′-deoxystrigol was identified in the root exudates of the hydroponically grown *Lotus japonicus* as the first substance to be characterized as a BF (Akiyama et al., 2005). Additionally, the natural strigolactones 5-deoxystrigol, sorgolactone, and strigol, as well as the synthetic strigolactone analog GR24, caused extensive AM hyphal branching in *Gigaspora margarita* at pg to ng levels, further supporting the idea that strigolactones work as branching factors (Akiyama et al., 2005). At present, more than 30 strigolactones have been characterized along with a few synthetic analogs like GR24 (Yoneyama and Brewer, 2021).

6.3 MECHANISM OF ACTION OF STRIGOLACTONES

Strigolactones cause AMF to branch; however, the exact mechanism causing this is still unknown. According to studies, BF extracted from carrot hairy roots caused *Gigaspora rosea* cells to develop more mitochondria, and their respiratory rate increased before they started to branch. Lipids, the primary form of carbon storage in AMF, are oxidatively catabolized as part of these crucial fungal reactions, and ATP is produced as a result. This suggests the stimulatory role of strigolactones in fungal respiration and lipid catabolism (Tamasloukht et al., 2003; Mitra et al., 2021c,d). This hypothesis was further supported by the work of Besserer et al. (2006), who recorded the positive effect of GR24 on the mitochondrial density, shape, distribution, and motility of *G. rosea*. However, the interaction between strigolactone and AMF doesn't stop there. Once the symbiosis is established between the AMF and the plant roots, the exudation of strigolactone is reduced (Lendzemo et al., 2007; Sun et al., 2008). This not only limits the further branching of the AMF but also suppresses the germination of parasitic weeds such as *Striga* and *Orobanche*. This property has huge agronomic implications as the AMF-plant root symbiosis and its impact on strigolactone production can be explored for effective management of parasitic weeds.

6.4 IMPORTANCE OF SYNTHETIC STRIGOLACTONES

Strigolactones play an important role in plant development, nutrition, and rhizosphere signaling (Borghi et al., 2021; Mitra et al., 2021e). However, the low amounts of strigolactones produced by plants, the short soil half-life, and the hydrolytic instability of strigolactones make it difficult to conduct research on strigolactones and their potential applications in agriculture. The amount of strigolactones produced was from 20 pg/plant in cotton to about 100 ng/g of roots in *Astragalus sinicus* (Sato et al., 2005; Yoneyama et al., 2011). The stability and half-life of these molecules vary in different environmental matrices as well, from a few hours in water to 6–8 days in acidic soil (Halouzka et al., 2018; Zwanenburg and Pospíšil, 2013). Thus, chemical synthesis of stable and effective strigolactone analogs will help enhance research on these chemicals and exploit their use in sustainable agricultural development. They can be useful for boosting the crop yield in nutrient-poor soil, especially nitrogen and phosphates, since the AMF strive better in nutrient-demanding conditions (Breuillin et al., 2010; Kretzschmar et al., 2012). Their stimulatory effect on the germination of parasitic weed seeds can be utilized to induce suicidal germination of these weeds and thereby control them (Yoneyama et al., 2019). These chemicals can be further used to stimulate mycorrhizal-induced resistance (MIR) against pests and diseases (Pozo and Azcón-Aguilar, 2007), as well as abiotic stresses, and to reduce the use of pesticides and fertilizers, taking a step toward sustainable agricultural practices.

6.5 STRUCTURAL FEATURES OF STRIGOLACTONES FOR BIOLOGICAL ACTIVITIES

Strigolactones are a class of terpenoid lactones with a distinctive structure made up of an enol-ether bond between a tricyclic (ABC ring) lactone and a methylbutenolide (D ring) (Figure 6.1). The strigolactones have several biological activities, including influencing the spore germination and hyphal branching of the AMF and stimulating the germination of parasitic weeds like *Striga* and *Orobanche*.

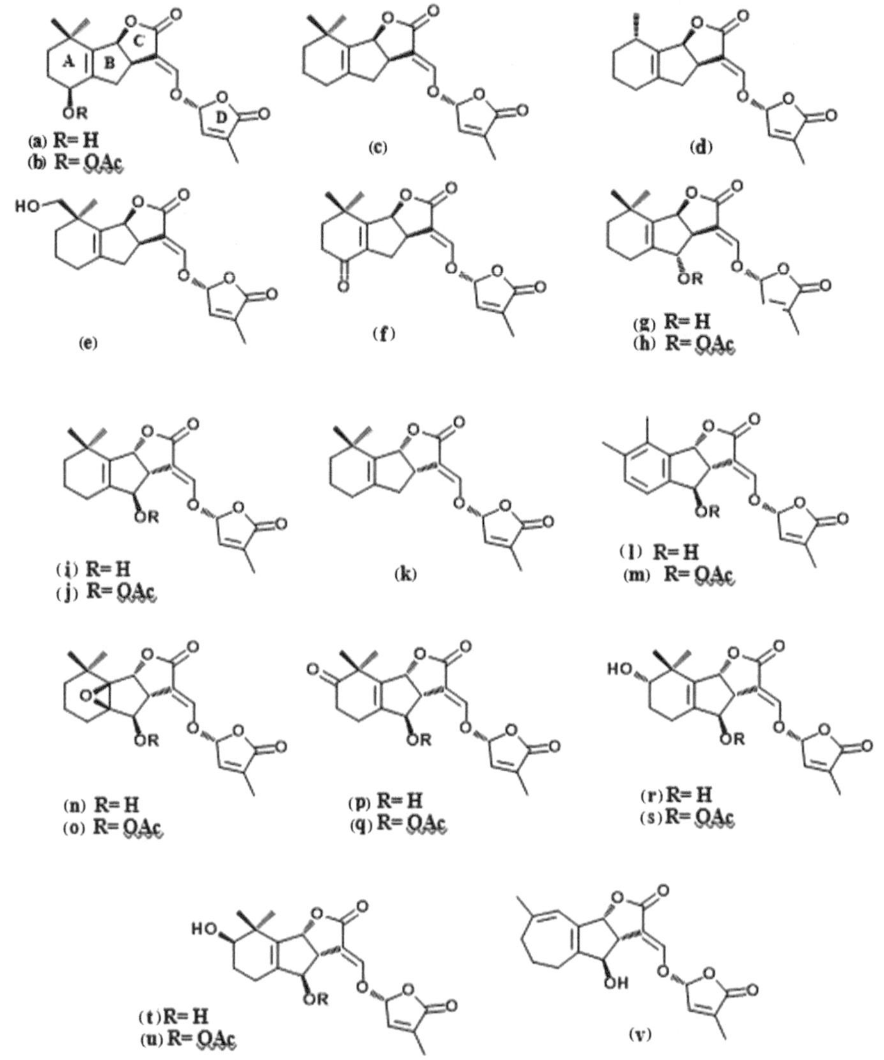

FIGURE 6.1 Canonical strigolactones. Strigol (a), strigyl acetate (b), 5-deoxystrigol (5DS) (c), sorgolactone (d), sorgomol (e), strigone (f), 4α-hydroxy-5-deoxystrigol (ent-2′-epi-orobanchol) (g), 4α-acetoxy-5-deoxystrigol (ent-2′-epi-orobanchyl acetate) (h), orobanchol (i), orobanchyl acetate (j), 4-deoxyorobanchol (4DO) (k), solanacol (l), solanacyl acetate (m), fabacol (n), fabacyl acetate (o), 7-oxoorobanchol (p), 7-oxoorobanchyl acetate (q), 7α-hydroxyorobanchol (r), 7α-hydroxyorobanchyl acetate (s), 7β-hydroxyorobanchol (t), 7β-hydroxyorobanchyl acetate (u), and medicaol (v). (Yoneyama et al., 2018.)

They also act as plant hormones and inhibit lateral shoot growth. These biological processes are all tightly linked to the compound's structure. Strigolactones' structure-activity relationship in terms of their germination stimulatory activity has been extensively studied (Hassanali, 1984; Mangnus et al., 1992). Mangnus and Zwanenburg (1992) identified the strigolactone molecule's C- and D rings as the crucial component for germination stimulation, and it was also demonstrated that the composition and configuration of the stimulants had a significant impact on the biological activity (Yoneyama et al., 2009; Zwanenburg et al., 2009). Contrarily, little research has been done on the structural characteristics of strigolactones, which are necessary to encourage the hyphal branching of the AMF. While the C-D ring is necessary for branching promotion, the A-B ring's presence and configuration are far more significant. Cleavage of the A and A-B rings of the tricyclic lactone ring significantly decreased the activity of hyphal branching in *Gigaspora margarita*. The biological action on the AMF was completely lost when the A-B ring from the strigolactone skeleton was removed, whereas the A ring shortened homolog GR7 was only marginally active. Hyphal branching in *Gigaspora rosea* showed a sharp decrease upon cleavage of the A ring (Besserer et al., 2006). Strigolactones' water breakdown products, formyl tricyclic ABC lactone and hydroxymethylbutenolide, failed to encourage hyphal branching, demonstrating that the loss of AMF activity is caused by the cleavage of the enol-ether link (Akiyama et al., 2010). A-B ring substitutions in strigolactones also had a dramatic influence on its activity. The activity was reduced when a hydroxyl group was added to the A ring of 5DS at either C-5 or C-9, as in strigol isomers and (+)-sorgomol. The amount of hyphal branching activity was significantly reduced when benzene was used in place of dimethylcyclohexene, as shown by 5DS between GR24. In a prior work (Akiyama et al., 2005), racemic sorgolactone demonstrated activity comparable to that of racemic 5DS, demonstrating that the presence of one methyl ring in the A ring is sufficient for the activity. Oxidation at either C-5 or C-7 (methylene to carbonyl) had no impact on the activity as well. Together, strigolactone activity in the AMF is decreased by polar substitution and ring replacement in A ring. Hydroxylation of the B ring at the C-4 position has a positive impact on hyphal branching. However, several studies have reported that non-canonical strigolactones (Figure 6.2) with an A and D ring and an enol-ether bond are still effective branching agents despite not having a B or C ring. Such strigolactones of the carlactone class have been discovered to promote the germination of parasitic weeds (Alder et al., 2012; Abe et al., 2014). Carlactone-like synthetic analogs have been synthesized, and studies have shown that they are exceptionally effective at promoting branching in AMF (Mori et al., 2016).

6.6 SYNTHETIC STRIGOLACTONES AND THEIR APPLICATION

Since the discovery of strigol in 1966, >14 additional strigolactones are reported in the root exudates of mono and dicot plants. But until recently, it was unknown how strigolactones in plants functioned. Since strigolactones are only present in

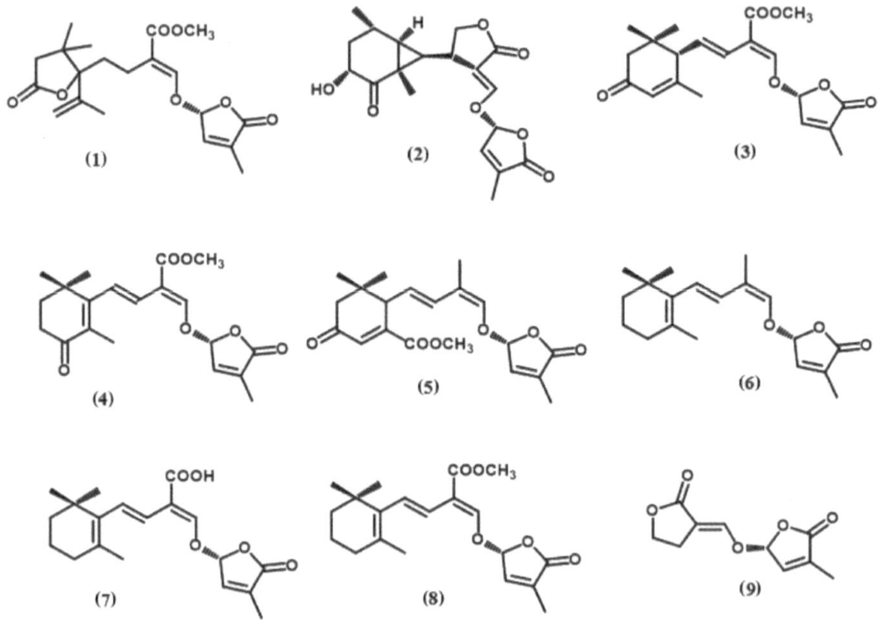

FIGURE 6.2 Non-canonical strigolactones. Zealactone (1), avenaol (2), heliolactone (3), carlactone (6), carlactonic acid (7) methyl carlactonoate (8), and GR 5 (9). (Yoneyama et al., 2018.)

very small quantities in plants, isolating them can be time-consuming and complex. Thus, an efficient synthesis of strigolactones is very much essential to study the functions and accurate mode of action of this novel class of phytohormones. Future research focused at avoiding issues with natural strigolactones, such as instability and limited abundance, may be made possible by the successful synthesis of novel strigolactones mimics. Several studies have shown interest in the finding of strigol analogs throughout the last few years. The synthesis of GR-24, which is still the gold standard for synthetic strigolactones and has incredibly potent biological action, was described by Johnson et al. (1981). They treated a ketene/ketene-iminium and an olefin via a [2+2] intramolecular cycloaddition and subsequent Baeyer-Villiger oxidation to obtain their tricyclic core structure. Zwanenburg et al. (2009) created synthetic strigolactones to explore the SAR and define their putative plant receptors. Lachia et al. (2012) developed a fast, effective, and supple stereoselective synthesis method for GR-24, which resulted in high yield and regioselectivity. The bioactivity of the synthesized strigolactone derivatives was evaluated both *in vitro* and *in vivo.*

To manufacture GR-24 analogs with extra substituents at specific places in the B and C rings for the germination of *Orobanche cumana* (broomrape) seeds, the method was further expanded by Lachia et al. (2014). The cyclobutanones undergo regioselective Baeyer-Villiger oxidation to yield the corresponding lactones. Following that, the germination activity of the nine GR-24 analogs was evaluated. The derivatives with an α-stereochemisty demonstrated strong seed germination stimulation. Kondo et al. (2007) reported that imino analogs with an oxime group in place of the enol-ether bridge strongly influenced the germination in *S. hermonthica* seeds. *Striga* germination has been shown to be significantly aided by carbamate analogs with outstanding femtomolar activity. These compounds replace the enol-ether bond (Uraguchi et al., 2018). A or B ring hydroxyl groups added to SLs have extremely strong effectiveness against parasitic weed germination (Brun et al., 2018). According to reports, the strigolactam analog of GR-24 is a better germination stimulant than GR-24 (Lachia et al., 2015). While the corresponding (–) epimers of GR-24 were inactive, the lactam analogs' (–) epimers are potent stimulators of seed germination.

However, limited research has been conducted to examine how synthetic strigolactones affect the branching of AMF. Borghi et al. (2021) studied the effect of GR-24 and other synthetic analogs on AMF branching. GR-24 along with its isomer SL1 promoted the branching significantly. The substitution with the N-acetyl group in the strigolactam moiety had a positive impact on hyphal branching. Akiyama et al. (2010) examined the ability of a number of chemically engineered strigolactones to induce hyphal branching in *Gigaspora margarita*. All of the examined compounds exhibited activity when the canonical strigolactone-like structure was present, but with varied patterns of activity. Tricyclic ABC lactones of strigolactones' A and A-B ring truncation significantly decreased hyphal branching activity. It has been demonstrated that hyphal branching depends on the interaction between the tricyclic lactone's C ring and the methylbutenolide D ring. However, it was discovered that the C-D part's bridge structure was replaceable with either alkoxy or imino ethers and was not always an enol ether. The activity was found to be strictly stereoselective and enantioselective (100 pg vs. 10 ng per disc). In general, the (+) isomers exhibited better hyphal branching than the (–) counterparts.

Non-canonical strigolactones and their analogs such as CL, CLA, MCLA and CLOH were chemically synthesized and tested to explore branching stimulatory activity in the AMF *Gigaspora margarita* (Mori et al., 2016). This study was carried out to test the hypothesis that carlactone along with its oxidized derivatives can act as host-derivative pre-colonization signals in AMF symbiosis. The extent of oxidation at C-19 methyl was found to be associated with hyphal branching activity. The natural canonical strigolactones strigol and sorgomol were equivalent in activity to carlactonoic acid (100 pg/disc), although carlacactone was only moderately active (100 ng/disc). Neither C-4 nor C-18 hydroxylation had a substantial impact on the activity. Formyl Meldrum's acid was combined with benzyl, cyclohexylmethyl, and cyclogeranyl alcohols (the A-ring portion) to create the ester and diester analogs of carlactone. The diester analogs were more active on the AMF (10 pg/disc) than ester analogs (1 ng to 100 pg/disc). These findings suggest the importance of the oxidation at C-19 (methyl to carboxyl) in carlactone over B-C ring formation in terms of activity on AMF.

6.7 CONCLUSION AND FUTURE PERSPECTIVE

The role of strigoactone in the rhizosphere is multidisciplinary, from supplying nutrients to alleviating biotic and abiotic stress, and from stimulating the germination of parasitic weeds to acting as a signaling molecule for AMF symbiosis; the applications are spread across a broad range. Much of its biological activity is achieved through AMF, where the strigolactone influences the germination as well as branching of the fungi. Chemical synthesis of natural strigolactone and strigolactone analogs opens new direction in exploring the bioactivity further as it helps to overcome the hurdles faced during working with natural strigolactones such as their minute quantity in root exudates and their stability in natural environment. Stereoselective synthesis further helps to explore the structure-activity relationships of strigolactones concerning their variety of biological activities and the discovery of other compounds with strigolactone-like activity. Improving the stability of these chemicals in natural environments like soil and water is still a challenge, and future work in this regard will further help in achieving sustainable development.

AUTHORS' CONTRIBUTIONS

BP, PS, SB, and PCM conceived the idea; PCM and SB collected the data and prepared the basic manuscript; BP and PS edited the manuscript; PS did some corrections and addition to the manuscript; BP incorporated the structures; SB and PCM revised and finalized the manuscript. All the authors read and approved the final manuscript.

CONFLICT OF INTEREST

The authors declare no conflict of interest.

REFERENCES

Abe, S., Sado, A., Tanaka, K., Kisugi, T., Asami, K., Ota, S., Kim, H.I., Yoneyama, K., Xie, X., Ohnishi, T. and Seto, Y., 2014. Carlactone is converted to carlactonoic acid by MAX1 in Arabidopsis and its methyl ester can directly interact with AtD14 in vitro. *Proceedings of the National Academy of Sciences, 111*(50), 18084–89. https://doi.org/10.1073/pnas.1410801111.

Akiyama, K., Matsuzaki, K.I. and Hayashi, H., 2005. Plant sesquiterpenes induce hyphal branching in arbuscular mycorrhizal fungi. *Nature, 435*(7043), 824–27. https://doi.org/10.1038/nature03608.

Akiyama, K., Ogasawara, S., Ito, S. and Hayashi, H., 2010. Structural requirements of strigolactones for hyphal branching in AM fungi. *Plant and Cell Physiology, 51*(7), 1104–17. https://doi.org/10.1093/pcp/pcq058.

Alder, A., Jamil, M., Marzorati, M., Bruno, M., Vermathen, M., Bigler, P., Ghisla, S., Bouwmeester, H., Beyer, P. and Al-Babili, S., 2012. The path from β-carotene to carlactone, a strigolactone-like plant hormone. *Science, 335*(6074), 1348–51. https://doi.org/10.1126/science.1218094.

Becard, G., Taylor, L.P., Douds, D.D., Pfeffer, P.E. and Doner, L.W., 1995. Flavonoids are not necessary plant signal compounds in arbuscular mycorrhizal symbioses. *MPMI-Molecular Plant Microbe Interactions*, *8*(2), 252–58.

Besserer, A., Puech-Pagès, V., Kiefer, P., Gomez-Roldan, V., Jauneau, A., Roy, S., Portais, J.C., Roux, C., Bécard, G. and Séjalon-Delmas, N., 2006. Strigolactones stimulate arbuscular mycorrhizal fungi by activating mitochondria. *PLoS Biology*, *4*(7), 226. https://doi.org/10.1371/journal.pbio.0040226.

Borghi, L., Screpanti, C., Lumbroso, A., Lachia, M., Gübeli, C. and De Mesmaeker, A., 2021. Efficiency and bioavailability of new synthetic strigolactone mimics with potential for sustainable agronomical applications. *Plant and Soil*, *465*(1), 109–23. https://doi.org/10.1007/s11104-021-04943-8.

Breuillin, F., Schramm, J., Hajirezaei, M., Ahkami, A., Favre, P., Druege, U., Hause, B., Bucher, M., Kretzschmar, T., Bossolini, E. and Kuhlemeier, C., 2010. Phosphate systemically inhibits development of arbuscular mycorrhiza in *Petunia hybrida* and represses genes involved in mycorrhizal functioning. *The Plant Journal*, *64*(6), 1002–17. https://doi.org/10.1111/j.1365-313X.2010.04385.x.

Brun, G., Braem, L., Thoiron, S., Gevaert, K., Goormachtig, S. and Delavault, P., 2018. Seed germination in parasitic plants: what insights can we expect from strigolactone research? *Journal of Experimental Botany*, *69*(9), 2265–80. https://doi.org/10.1093/jxb/erx472.

Buee, M., Rossignol, M., Jauneau, A., Ranjeva, R. and Bécard, G., 2000. The presymbiotic growth of arbuscular mycorrhizal fungi is induced by a branching factor partially purified from plant root exudates. *Molecular Plant-Microbe Interactions*, *13*(6), 693–98. https://doi.org/10.1094/MPMI.2000.13.6.693.

Halouzka, R., Tarkowski, P., Zwanenburg, B. and Ćavar Zeljković, S., 2018. Stability of strigolactone analog GR24 toward nucleophiles. *Pest Management Science*, *74*(4), 896–904. https://doi.org/10.1002/ps.4782.

Harrison, M.J., 2005. Signaling in the arbuscular mycorrhizal symbiosis. *Annual Review of Microbiology*, *59*, 19–42. https://doi.org/10.1146/annurev.micro.58.030603.123749.

Hassanali, A., 1984. Strigol analogues: synthetic achievements and prospects. In: *Striga--Biology and Control*, edited by Ayensu, E.S., Dogget, H., Keynes, K.D., Marton-Lefevre, J., Musselman, L.J., Parker, C., et al., ICSU Press, Paris, France, pp. 125–32.

Johnson, A.W., Gowada, G., Hassanali, A., Knox, J., Monaco, S., Razavi, Z. and Rosebery, G., 1981. The preparation of synthetic analogues of strigol. *Journal of the Chemical Society, Perkin Transactions 1*, 1734–43. https://doi.org/10.1039/P19810001734.

Kondo, Y., Tadokoro, E., Matsuura, M., Iwasaki, K., Sugimoto, Y., Miyake, H., Takikawa, H. and Sasaki, M., 2007. Synthesis and seed germination stimulating activity of some imino analogs of strigolactones. *Bioscience, Biotechnology and Biochemistry*, 0710040585. https://doi.org/10.1271/bbb.70398.

Kretzschmar, T., Kohlen, W., Sasse, J., Borghi, L., Schlegel, M., Bachelier, J.B., Reinhardt, D., Bours, R., Bouwmeester, H.J. and Martinoia, E., 2012. A petunia ABC protein controls strigolactone-dependent symbiotic signalling and branching. *Nature*, *483*(7389), 341–44. https://doi.org/10.1038/nature10873.

Lachia, M., Jung, P.M. and De Mesmaeker, A., 2012. A novel approach toward the synthesis of strigolactones through intramolecular [2+2] cycloaddition of ketenes and ketene-iminiums to olefins. Application to the asymmetric synthesis of GR-24. *Tetrahedron Letters*, *53*(34), 4514–17. https://doi.org/10.1016/j.tetlet.2012.06.013.

Lachia, M., Wolf, H.C. and De Mesmaeker, A., 2014. Synthesis of strigolactones analogues by intramolecular [2+2] cycloaddition of ketene-iminium salts to olefins and their activity on *Orobanche cumana* seeds. *Bioorganic & Medicinal Chemistry Letters*, *24*(9), 2123–28. https://doi.org/10.1016/j.bmcl.2014.03.044.

Lachia, M., Wolf, H.C., Jung, P.J.M., Screpanti, C. and De Mesmaeker, A., 2015. Strigolactam: new potent strigolactone analogues for the germination of *Orobanche cumana*. *Bioorganic & Medicinal Chemistry Letters*, *25*(10), 2184–88. https://doi.org/10.1016/j.bmcl.2015.03.056.

Lendzemo, V.W., Kuyper, T.W., Matusova, R., Bouwmeester, H.J. and Ast, A.V., 2007. Colonization by arbuscular mycorrhizal fungi of sorghum leads to reduced germination and subsequent attachment and emergence of *Striga hermonthica*. *Plant Signaling & Behavior*, *2*(1), 58–62. https://doi.org/10.4161/psb.2.1.3884.

Mangnus, E.M., Van Vliet, L.A., Vandenput, D.A. and Zwanenburg, B., 1992. Structural modifications of strigol analogs. Influence of the B and C rings on the bioactivity of the germination stimulant GR24. *Journal of Agricultural and Food Chemistry*, *40*(7), 1222–29. https://doi.org/10.1021/jf00019a030.

Mangnus, E.M. and Zwanenburg, B., 1992. Synthesis and biological evaluation of A-ring analogs of the natural germination stimulant strigol. *Recueil des Travaux Chimiques des Pays-Bas*, *111*(3), 155–9. https://doi.org/10.1002/recl.19921110308.

Mitra, D., Be, G.S., Khoshru, B., De Los Santos Villalobos, S., Belz, C., Chaudhary, P., Shahri, F.N., Djebaili, R., Adeyemi, N.O., El-Ballat, E.M. and El-Esawi, M.A., 2021a. Impacts of arbuscular mycorrhizal fungi on rice growth, development, and stress management with a particular emphasis on strigolactone effects on root development. *Communications in Soil Science and Plant Analysis*, *52*(14), 1591–621.

Mitra, D., Djebaili, R., Pellegrini, M., Mahakur, B., Sarker, A., Chaudhary, P., Khoshru, B., Gallo, M.D., Kitouni, M., Barik, D.P. and Panneerselvam, P., 2021b. Arbuscular mycorrhizal symbiosis: plant growth improvement and induction of resistance under stressful conditions. *Journal of Plant Nutrition*, *44*(13), 1993–2028.

Mitra, D., Mondal, R., Khoshru, B., Shadangi, S., Mohapatra, P.K.D. and Panneerselvam, P., 2021d. Rhizobacteria mediated seed bio-priming triggers the resistance and plant growth for sustainable crop production. *Current Research in Microbial Sciences*, *2*, 100071.

Mitra, D., Panneerselvam, P., Parameswaran, C. and Mohapatra, P.K.D. 2021e. Is strigolactone (SL) hormone secretion by rice plants is essential for improving mycorrhizal fungal association? *National Conference of Plant Physiology - 2021 Frontiers of Plant Physiology for Climate Smart Agriculture*, Baramati, Pune Affiliation: National Institute of Abiotic Stress Management.

Mitra, D., Rad, K.V., Chaudhary, P., Ruparelia, J., Sagarika, M.S., Boutaj, H., Mohapatra, P.K.D. and Panneerselvam, P., 2021c. Involvement of strigolactone hormone in root development, influence and interaction with mycorrhizal fungi in plant: mini-review. *Current Research in Microbial Sciences*, *2*, 100026.

Mori, N., Nishiuma, K., Sugiyama, T., Hayashi, H. and Akiyama, K., 2016. Carlactone-type strigolactones and their synthetic analogues as inducers of hyphal branching in arbuscular mycorrhizal fungi. *Phytochemistry*, *130*, 90–8. https://doi.org/10.1016/j.phytochem.2016.05.012.

Nagahashi, G. and Douds, D.D., 1999. Rapid and sensitive bioassay to study signals between root exudates and arbuscular mycorrhizal fungi. *Biotechnology Techniques*, *13*(12), 893–97. https://doi.org/10.1023/A:1008938527757.

Pozo, M.J. and Azcón-Aguilar, C., 2007. Unraveling mycorrhiza-induced resistance. *Current Opinion in Plant Biology*, *10*(4), 393–98. https://doi.org/10.1016/j.pbi.2007.05.004.

Sato, D., Awad, A.A., Takeuchi, Y. and Yoneyama, K., 2005. Confirmation and quantification of strigolactones, germination stimulants for root parasitic plants Striga and Orobanche, produced by cotton. *Bioscience, Biotechnology and Biochemistry*, *69*(1), 98–102. https://doi.org/10.1271/bbb.69.98.

Simon, L., Bousquet, J., Lévesque, R.C. and Lalonde, M., 1993. Origin and diversification of endomycorrhizal fungi and coincidence with vascular land plants. *Nature*, *363*(6424), 67–9. https://doi.org/10.1038/363067a0.

Sun, Y., Li, X.L. and Feng, G., 2008. Effect of arbuscular mycorrhizal colonization on ecological functional traits of ephemerals in the Gurbantonggut desert. *Symbiosis* *46*, 121–27.

Tamasloukht, M.B., Séjalon-Delmas, N., Kluever, A., Jauneau, A., Roux, C., Bécard, G. and Franken, P., 2003. Root factors induce mitochondrial-related gene expression and fungal respiration during the developmental switch from asymbiosis to presymbiosis in the arbuscular mycorrhizal fungus Gigaspora rosea. *Plant Physiology*, *131*(3), 1468–78. https://doi.org/10.1104/pp.012898.

Uraguchi, D., Kuwata, K., Hijikata, Y., Yamaguchi, R., Imaizumi, H., Am, S., Rakers, C., Mori, N., Akiyama, K., Irle, S. and McCourt, P., 2018. A femtomolar-range suicide germination stimulant for the parasitic plant *Striga hermonthica*. *Science*, *362*(6420), 1301–05. https://doi.org/10.1126/science.aau5445.

Yoneyama, K. and Brewer, P.B., 2021. Strigolactones, how are they synthesized to regulate plant growth and development? *Current Opinion in Plant Biology*, *63*, 102072. https://doi.org/10.1016/j.pbi.2021.102072.

Yoneyama, K., Xie, X., Kisugi, T., Nomura, T., Sekimoto, H., Yokota, T. and Yoneyama, K., 2011. Characterization of strigolactones exuded by Asteraceae plants. *Plant Growth Regulation*, *65*(3), 495–504. https://doi.org/10.1007/s10725-011-9620-z.

Yoneyama, K., Xie, X., Yoneyama, K., Kisugi, T., Nomura, T., Nakatani, Y., Akiyama, K. and McErlean, C.S., 2018. Which are the major players, canonical or non-canonical strigolactones? *Journal of Experimental Botany*, *69*(9), 2231–39. https://doi.org/10.1093/jxb/ery090.

Yoneyama, K., Xie, X., Yoneyama, K., Nomura, T., Takahashi, I., Asami, T., Mori, N., Akiyama, K., Kusajima, M. and Nakashita, H., 2019. Regulation of biosynthesis, perception, and functions of strigolactones for promoting arbuscular mycorrhizal symbiosis and managing root parasitic weeds. *Pest Management Science*, *75*(*9*), 2353–59. https://doi.org/10.1016/j.pbi.2021.102072.

Yoneyama, K., Xie, X., Yoneyama, K. and Takeuchi, Y., 2009. Strigolactones: structures and biological activities. *Pest Management Science: Formerly Pesticide Science*, *65*(5), 467–70. https://doi.org/10.1002/ps.1726.

Zwanenburg, B. and Pospíšil, T., 2013. Structure and activity of strigolactones: new plant hormones with a rich future. *Molecular Plant*, *6*(1), 38–62. https://doi.org/10.1093/mp/sss141.

Zwanenburg, B., Mwakaboko, A.S., Reizelman, A., Anilkumar, G. and Sethumadhavan, D., 2009. Structure and function of natural and synthetic signalling molecules in parasitic weed germination. *Pest Management Science: Formerly Pesticide Science*, *65*(5), 478–91. https://doi.org/10.1002/ps.1706.

7 The Beneficial Role of Arbuscular Mycorrhizal Fungi and Their Associated Bacteria for Plant Growth Promotion and Nutrient Management in Rice Cultivation

Bahman Khoshru
University of Tabriz

Debasis Mitra
ICAR-National Rice Research Institute

CONTENTS

7.1 INTRODUCTION

Rice (*Oryza sativa* L.) is known as a monocotyledon model plant and as one of the most important agricultural products (Sawers et al., 2008; Wang et al., 2015). In the rhizosphere of this plant, a narrow area of soil that includes the root framework, there are many interactions between the plant root and microbes (Ding et al., 2019). Arbuscular mycorrhizal (AM) fungi are one of the most important symbionts of plants (Mitra et al., 2021; Panneerselvam et al., 2017; Sahoo et al., 2017). But in the case of the rice plant, which is mostly cultivated in the wetland

DOI: 10.1201/9781003354086-7

system and has anaerobic environments (FAO, 2018), it is thought that AMF will have problems in establishing symbiosis with rice. There are conflicting results for AMF studies in rice plants. Some studies have reported that AMF are rare or absent in rice roots in flooded paddy fields (Ilag et al., 1987; Lumini et al., 2011; Panneerselvam et al., 2019, 2020). In other studies, rice root colonization in paddy fields by AMF has been reported (Wang et al., 2015; Bernaola et al., 2018). In general, flooding has an inhibitory effect on the formation of AMF symbiosis with rice roots (Zhang et al., 2015). It has been shown that the formation of AMF symbiosis in the rice plant, although it is absent or rare in the early stages of growth, happens in the heading and ripening stages of the plant, which is due to the formation and development of aerenchyma and the transfer of oxygen to the roots (Wang et al., 2015). The effect of AMF inoculation on rice plant growth under conditions of submerged cultivation (Solaiman and Hirata, 1995; Bernaola et al., 2018) and non-flooded cultivation (Ruiz Sánchez et al., 2015; Zhang et al., 2015) is reported.

The area that is affected by both roots and mycorrhizal fungi is known as the mycorrhizosphere, which includes a more specific space called "hyposphere" that refers to the area around fungal hyphae (Rambelli, 1973; Priyadharsini et al., 2016). In this narrow area, due to the secretion of organic substances from the roots and mycorrhiza, it has high microbial activity (Linderman, 1991; Frey-Klett et al., 2005). In the "mycorrhizosphere," there are different types of microbial groups (beneficial or harmful) that can affect the interaction of AMF with the plant, such as the degree of colonization and reproduction of the fungus (Hoeksema et al., 2010). Therefore, the mycorrhizal response of rice can be affected by the change of related microbial communities (positive or negative) (Ordoñez et al., 2016). The response of rice plant growth in interaction with AMF is the result of the balance between the beneficial and negative effects of the mentioned microbial communities (Van der Putten et al., 2013). Bacterial communities are dominant in the rice rhizosphere and have a great impact on plant growth (Breidenbach et al., 2016). These communities include mycorrhizal helping bacteria (MHB) (Mitra et al., 2020; Saritha et al., 2021; Panneerselvam et al., 2017), endobacteria, plant growth-promoting rhizobacteria (PGPR) (Panneerselvam et al., 2012), and deleterious bacteria (DB) (Miransari, 2011; Zhang et al., 2021). Understanding the interactions between plant-AMF-bacteria can lead to strengthening the efficiency of this tripartite symbiosis (Figure 7.1).

In this chapter, there will be an overview of the positive effects of the potential application of AMF in paddy fields, and the mechanisms involved in the interaction between rice plants and AMF in relation to the different types of microbial communities in the rice rhizosphere will be discussed. In this chapter, after discussing mycorrhizosphere microbial diversity and their interactions with plant-AMF symbiosis, an outlook on future research priorities is presented, and the potential application of AMF to improve rice plant performance is emphasized (Table 7.1).

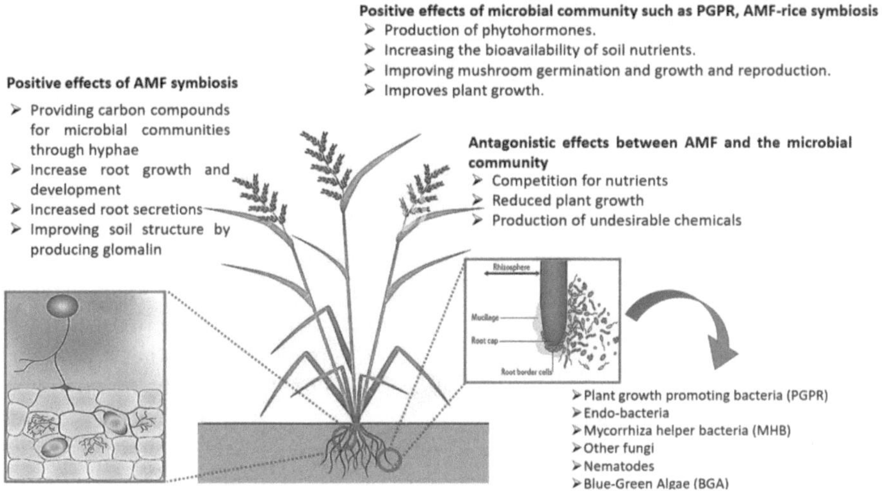

FIGURE 7.1 The interactions between AMF and different microbial communities of the rice rhizosphere.

TABLE 7.1
Some of the Major Bacteria Interacting with AMF in Rice Rhizosphere

Bacteria	AMF	Rice plant response	References
S. thermocarboxydus	*F. mosseae*	Development of plant growth	Lasudee et al. (2018)
P. jessenii	A natural AMF consortium (Mnat)	No effect	Artursson et al. (2006)
P. synxantha	Single-spore AMF strain (Mss2)	No effect	Mäder et al. (2011)
P. fluorescens	*R. irregularis*	Increase grain yield	Xiao et al. (2020)
P. putida	*R. irregularis*	Increasing yield and reducing salinity	Norouzinia et al. (2020)
H. seropedicae	*F. mosseae*	Increase plant growth	Hoseinzade et al. (2016)
A. brasilense	*F. mosseae*	Rice growth grain yield	Bao et al. (2022)
B. cepacia	*F. mosseae*	P uptake	Bonfante e and Desirò (2003)
A. Lipoferum	*Glomus* sp. and *Gigasporasp* sp.	Increase plant growth	Primieri et al. (2022)
B. megaterium	*Acaulospora* sp.	Increase grain yield	Prem Kumari and Prabina (2017)

7.2 CONCLUSION

The characteristics of soil structure and the growth and development of rice roots can be significantly influenced by AMF and the interactions of different microbial types of rice mycorrhizosphere. Different microbial groups interact with AMF, including non-mycorrhizal fungi, bacteria (PGPR, MHB, endophytes, etc.), and other soil fauna. The exchange of nutrients and other compounds is essential in the ternary symbiosis of rice plants, AMF, and other microbes. Each of these three living parties maintains its contribution to preserve this triple symbiosis, for example, the root of the rice plant secretes compounds such as polysaccharides, organic acids, amino acids, and hormones. AMF use its hyphae to access more water and nutrients and provide plant photosynthetic products to other microorganisms. Microbes also produce compounds such as phytohormones, polysaccharides, siderophores, organic acids, and antibiotics. In the rice-AMF symbiosis, several pathways and processes are associated with microbial mediation. Based on the above, we are only at the beginning of understanding the complexity of the interaction between AMF and microbes in the rice mycorrhizosphere. Most of the studies have been done on the interactions of AMF with rice plant, while more studies should be done on the interactions of other microbes and their impact in this multiple symbiosis. Determining the most efficient microbe in symbiosis with AMF can maximize the benefits of this triple symbiosis. The limitation of microbial culture in the conventional method prevents the identification and understanding of the types of microbes related to AMF and plant interactions; therefore, the use of techniques such as metagenomics and proteomics to investigate the mycorrhizosphere of rice will lead to very interesting and useful results.

REFERENCES

Artursson, V., Finlay, R.D. and Jansson, J.K., 2006. Interactions between arbuscular mycorrhizal fungi and bacteria and their potential for stimulating plant growth. *Environmental Microbiology*, *8*(1): 1–10.

Bao, X., Zou, J., Zhang, B., Wu, L., Yang, T. and Huang, Q., 2022. Arbuscular mycorrhizal fungi and microbes interaction in rice mycorrhizosphere. *Agronomy*, *12*(6): 1277.

Bernaola, L., Cange, G., Way, M.O., Gore, J., Hardke, J. and Stout, M., 2018. Natural colonization of rice by arbuscular mycorrhizal fungi in different production areas. *Rice Science*, *25*(3): 169–174.

Bonfante, P. and Desirò, A., 2017. Who lives in a fungus? The diversity, origins and functions of fungal endobacteria living in Mucoromycota. *The ISME Journal*, *11*(8): 1727–1735.

Breidenbach, B., Pump, J. and Dumont, M.G., 2016. Microbial community structure in the rhizosphere of rice plants. *Frontiers in Microbiology*, *6:* 1537.

Ding, L.J., Cui, H.L., Nie, S.A., Long, X.E., Duan, G.L. and Zhu, Y.G., 2019. Microbiomes inhabiting rice roots and rhizosphere. *FEMS Microbiology Ecology*, *95*(5): fiz040.

FAO. 2018. Rice Market Monitor. Available online: https://www.fao.org/3/I9243EN/i9243en.pdf (accessed on 25 April 2021).

Frey-Klett, P., Chavatte, M., Clausse, M.L., Courrier, S., Roux, C.L., Raaijmakers, J., Martinotti, M.G., Pierrat, J.C. and Garbaye, J., 2005. Ectomycorrhizal symbiosis affects functional diversity of rhizosphere fluorescent pseudomonads. *New Phytologist*, *165*(1): 317–328.

Hoeksema, J.D., Chaudhary, V.B., Gehring, C.A., Johnson, N.C., Karst, J., Koide, R.T., Pringle, A., Zabinski, C., Bever, J.D., Moore, J.C. and Wilson, G.W., 2010. A meta-analysis of context-dependency in plant response to inoculation with mycorrhizal fungi. *Ecology Letters*, *13*(3): 394–407.

Hoseinzade, H., Ardakani, M.R., Shahdi, A., Rahmani, H.A., Noormohammadi, G. and Miransari, M., 2016. Rice (*Oryza sativa* L.) nutrient management using mycorrhizal fungi and endophytic Herbaspirillum seropedicae. *Journal of Integrative Agriculture*, *15*(6): 1385–1394.

Ilag, L.L., Rosales, A.M., Elazegui, F.A. and Mew, T.W., 1987. Changes in the population of infective endomycorrhizal fungi in a rice-based cropping system. *Plant and Soil*, *103*(1): 67–73.

Lasudee, K., Tokuyama, S., Lumyong, S. and Pathom-Aree, W., 2018. Actinobacteria associated with arbuscular mycorrhizal Funneliformis mosseae spores, taxonomic characterization and their beneficial traits to plants: Evidence obtained from mung bean (Vigna radiata) and Thai jasmine rice (*Oryza sativa*). *Frontiers in Microbiology*, *9:* 1247.

Linderman, R.G., 1991. Mycorrhizal interactions in the rhizosphere. In: D.L. Keister and P.B. Cregan (Eds.), *The Rhizosphere and Plant Growth* (pp. 343–348). Springer, Dordrecht.

Lumini, E., Vallino, M., Alguacil, M.M., Romani, M. and Bianciotto, V., 2011. Different farming and water regimes in Italian rice fields affect arbuscular mycorrhizal fungal soil communities. *Ecological Applications*, *21*(5): 1696–1707.

Mäder, P., Kaiser, F., Adholeya, A., Singh, R., Uppal, H.S., Sharma, A.K., Srivastava, R., Sahai, V., Aragno, M., Wiemken, A. and Johri, B.N., 2011. Inoculation of root microorganisms for sustainable wheat–rice and wheat–black gram rotations in India. *Soil Biology and Biochemistry*, *43*(3): 609–619.

Miransari, M., 2011. Interactions between arbuscular mycorrhizal fungi and soil bacteria. *Applied Microbiology and Biotechnology*, *89*(4): 917–930.

Mitra, D., Djebaili, R., Pellegrini, M., Mahakur, B., Sarker, A., Chaudhary, P., Khoshru, B., Gallo, M.D., Kitouni, M., Barik, D.P. and Panneerselvam, P., 2021. Arbuscular mycorrhizal symbiosis: plant growth improvement and induction of resistance under stressful conditions. *Journal of Plant Nutrition*, *44*(13): 1993–2028.

Mitra, D., Uniyal, N., Panneerselvam, P., Senapati, A. and Ganeshamurthy, A.N., 2020. Role of mycorrhiza and its associated bacteria on plant growth promotion and nutrient management in sustainable agriculture. *International Journal of Life Sciences and Applied Sciences*, *1*(1): 1–1.

Norouzinia, F., Ansari, M.H., Aminpanah, H. and Firozi, S., 2020. Alleviation of soil salinity on physiological and agronomic traits of rice cultivars using Arbuscular mycorrhizal fungi and Pseudomonas strains under field conditions. *Revista de Agricultura Neotropical*, *7*(1): 25–42.

Ordoñez, Y.M., Fernandez, B.R., Lara, L.S., Rodriguez, A., Uribe-Velez, D. and Sanders, I.R., 2016. Bacteria with phosphate solubilizing capacity alter mycorrhizal fungal growth both inside and outside the root and in the presence of native microbial communities. *PLoS One*, *11*(6): 54438.

Panneerselvam, P., Kumar, U., Senapati, A., Parameswaran, C., Anandan, A., Kumar, A., Jahan, A., Padhy, S.R. and Nayak, A.K., 2020. Influence of elevated CO_2 on arbuscular mycorrhizal fungal community elucidated using Illumina MiSeq platform in sub-humid tropical paddy soil. *Applied Soil Ecology*, *145:* 103344.

Panneerselvam, P., Kumar, U., Sugitha, T.C.K., Parameswaran, C., Sahoo, S., Binodh, A.K., Jahan, A. and Anandan, A., 2017. Arbuscular mycorrhizal fungi (AMF) for sustainable rice production. In: T. K. Adhya, B. Lal, B. Mohapatra, D. Paul, and S. Das (Eds.), *Advances in Soil Microbiology: Recent Trends and Future Prospects* (pp. 99–126). Springer, Singapore.

Panneerselvam, P., Mohandas, S., Saritha, B., Upreti, K.K., Poovarasan, M.A. and Sulladmath, V.V., 2012. Glomus mosseae associated bacteria and their influence on stimulation of mycorrhizal colonization, sporulation, and growth promotion in guava (*Psidium guajava* L.) seedlings. *Biological Agriculture & Horticulture, 28*(4): 267–279.

Panneerselvam, P., Sahoo, S., Senapati, A., Kumar, U., Mitra, D., Parameswaran, C., Anandan, A., Kumar, A., Jahan, A. and Nayak, A.K., 2019. Understanding interaction effect of arbuscular mycorrhizal fungi in rice under elevated carbon dioxide conditions. *Journal of Basic Microbiology, 59*(12): 1217–1228.

Prem Kumari, S.M. and Prabina, B.J., 2017. Impact of the mixed consortium of indigenous arbuscular mycorrhizal fungi (AMF) on the growth and yield of rice (*ORYZA SATIVA* L.) under the system of rice intensification (SRI). *International Journal of Environment, Agriculture and Biotechnology, 2*(2): 238743.

Primieri, S., Magnoli, S.M., Koffel, T., Stürmer, S.L. and Bever, J.D., 2022. Perennial, but not annual legumes synergistically benefit from infection with arbuscular mycorrhizal fungi and rhizobia: a meta-analysis. *New Phytologist, 233*(1): 505–514.

Priyadharsini, P., Rojamala, K., Ravi, R.K., Muthuraja, R., Nagaraj, K. and Muthukumar, T., 2016. Mycorrhizosphere: The extended rhizosphere and its significance. In: D.K. Choudhary, A. Varma, and N. Tuteja (Eds.), *Plant-Microbe Interaction: An Approach to Sustainable Agriculture* (pp. 97–124). Springer, Singapore.

Rambelli, A.N.G.E.L.O., 1973. The rhizosphere of mycorrhizae. In: G.C. Marks and T.T. Kozlowski (Eds.), *Ectomycorrhizae: Their Ecology and Physiology* (pp. 299–349). Academic Press, New York.

Ruiz Sánchez, M., Santana Baños, Y., Muñoz Hernández, Y., Yoan Martínez, A., Benitez, M., Vishnu Bharat, B. and Peña Chávez, Y., 2015. Arbuscular mycorrhizal symbiosis in rice plants in flooded and no flooded conditions. *Acta Agronómica, 64*(3): 227–233.

Sahoo, S., Panneerselvam, P., Chowdhury, T., Kumar, A., Kumar, U., Jahan, A., Senapati, A. and Anandan, A., 2017. Understanding the AM fungal association in flooded rice under elevated CO2 condition. *ORYZA-An International Journal on Rice, 54*(3): 290–297.

Saritha, B., Panneerselvam, P., Srinivas, K., Debasis, M. and Ansuman, S., 2021. Enhancing the sapota [Manilkara achras (Mill) Forsberg] yield through intervention of Arbuscular mycorrhizal fungi and its associated bacteria. *Research Journal of Biotechnology, 16:* 99–109.

Sawers, R.J., Gutjahr, C. and Paszkowski, U., 2008. Cereal mycorrhiza: an ancient symbiosis in modern agriculture. *Trends in Plant Science, 13*(2): 93–97.

Solaiman, M.Z. and Hirata, H., 1995. Effects of indigenous arbuscular mycorrhizal fungi in paddy fields on rice growth and N, P, K nutrition under different water regimes. *Soil Science and Plant Nutrition, 41*(3): 505–514.

Van der Putten, W.H., Bardgett, R.D., Bever, J.D., Bezemer, T.M., Casper, B.B., Fukami, T., Kardol, P., Klironomos, J.N., Kulmatiski, A., Schweitzer, J.A. and Suding, K.N., 2013. Plant–soil feedbacks: the past, the present and future challenges. *Journal of Ecology, 101*(2): 265–276.

Wang, Y., Li, T., Li, Y., Björn, L.O., Rosendahl, S., Olsson, P.A., Li, S. and Fu, X., 2015. Community dynamics of arbuscular mycorrhizal fungi in high-input and intensively irrigated rice cultivation systems. *Applied and Environmental Microbiology*, *81*(8): 2958–2965.

Xiao, A.W., Li, Z., Li, W.C. and Ye, Z.H., 2020. The effect of plant growth-promoting rhizobacteria (PGPR) on arsenic accumulation and the growth of rice plants (*Oryza sativa* L.). *Chemosphere*, *242*:125136.

Zhang, L., Zhou, J., George, T.S., Limpens, E. and Feng, G., 2021. Arbuscular mycorrhizal fungi conducting the hyphosphere bacterial orchestra. *Trends in Plant Science*, 27(4): 402–411.

Zhang, S., Wang, L., Ma, F., Bloomfield, K.J., Yang, J. and Atkin, O.K., 2015. Is resource allocation and grain yield of rice altered by inoculation with arbuscular mycorrhizal fungi? *Journal of Plant Ecology*, *8*(4): 436–448.

8 An Insight of Physiological and Molecular Mechanisms of Arbuscular Mycorrhizal Fungi – Rice Symbiosis in Stress Alleviation

Manju Chaithra
ICAR-Indian Agricultural Research Institute

Amit Kumar Dutta and Mahwish Firdous
Amity University

Debasis Mitra
Raiganj University

CONTENTS

8.1 ARBUSCULAR MYCORRHIZAL FUNGI (AMF)

Arbuscular mycorrhizal fungi (AMF) are a type of mycorrhiza in which the symbiont fungus penetrates the cortical cells of the roots of a vascular plant forming arbuscules. AMF are soil microorganisms able to form mutualistic symbiosis with most terrestrial plants. Spores that are present in the soil germinate, infect the root system, and form arbuscule structures inside the cells. AMF, being natural root symbionts, provide essential plant inorganic nutrients to host plants,

DOI: 10.1201/9781003354086-8

thereby improving growth and yield under unstressed and stressed regimes. The role of AMF as a biofertilizer can potentially strengthen plants' adaptability to changing environment (Panneerselvam et al., 2017, 2012).

Arbuscular mycorrhizal fungi are categorized in a monogenetic class, the phylum Glomeromycota (Stürmer, 2012). They fulfill their life by consuming plant products. The colonization of AMF can increase the absorption of mineral nutrients in the plant, largely nitrogen and phosphorus in the interchange of carbon (Smith and Read, 2008). AMF are also known as biofertilizers (Begum et al., 2019; Panneerselvam et al., 2013). AMF can assist the host plant in the upregulation of the sufferance process and halt the downregulation of key metabolic pathway. AMF can ease the host plant to grow under stressful condition.

8.2 RESPONSE OF PLANT UNDER STRESS CONDITION

Drought stress is responsible for the deterioration of the soil and presents severe threats to agriculture worldwide. Most of the research over the last few years has focused on symbiotic mechanisms of AMF for protecting plants against drought stress, which verdict, symbiosis often marks an increased accumulation of osmoregulators, nutrient uptake, WUE, and photosynthetic rate (Zhao et al., 2015). From a research viewpoint, efforts from both AMF and drought stress fields have improved our understanding of these mechanisms. In this regard, more recently, it was confirmed that AM symbiosis-specific downstream responses control a combination of morphological, biochemical, and physiological plant characteristics (Mitra et al., 2020, 2022).

AMF establish symbiotic interaction with 80% of known land plants. It has a pronounced impact on plant growth, water absorption, mineral nutrition, and protection from abiotic stresses (Mitra et al., 2021a,c). Plants are very dynamic systems having great adaptability under continuously changing drying conditions. In this regard, the function of AMF as a biological tool for improving plant drought stress tolerance and phenotypic plasticity, in terms of establishing mutualistic associations, seems an innovative approach toward sustainable agriculture. However, a better understanding of these complex interconnected signaling pathways and AMF-mediated mechanisms that regulate the drought tolerance in plants will enhance its potential application as an innovative approach in environmentally friendly agriculture (Panneerselvam et al., 2020, 2019).

It is extensively believed that soil salinization is the biggest problem to the global food security. The meaning of soil salinization is to abolish the growth of plant by altering its net assimilation rate and vegetative growth following decreases in productivity by 20% worldwide (Ahanger et al., 2017). It also increases the growth of oxygen reactive species (Herbinger et al., 2002; Ahanger and Agarwal, 2017; Ahanger et al., 2017, 2018). It is observed that the cautious use of AMF can control the salinity stress (Santander et al., 2019). For example, influx/efflux of ion, translocation of sodium ion from root to shoot, and compartmentalization have been controlled by the AMF inoculation. Better resistance in the form of greater yield of glycine betaine, soluble sugar, and proline can also be observed against salinity stress (Garg and Baher, 2013).

In soil, fungi can be characterized by their nutritional strategy: most fungi are saprotrophs obtaining nutrients from dead organic matter, while a small group of parasitic or mutualistic fungi feed on living organisms (Saritha et al., 2021). Although some fungi are always in one category, others can switch among the saprotrophic, parasitic, and mutualistic lifestyles depending on circumstances. In natural ecosystems, mycorrhizal fungi can be found as free-living organisms but need to associate with plant roots to complete their life cycle. This association is categorized as mutualistic because the fungus improves plant nutrient acquisition, water absorption, growth, and resistance to biotic and abiotic stresses, while the host plant provides carbohydrates and nutrients to the fungus allowing its growth and development (Bonfante and Genre, 2010; Mitra et al., 2021b).

Mycorrhizal fungi emerged with plant colonization of land around 450 million years ago (Remy et al., 1994; Taylor et al., 1995). Coincidentally, the earliest association of plant with fungus was dated to 407 million years ago, suggesting that ancestral plant-fungus symbiosis may have facilitated plant transition from aquatic to continental environments (Strullu-Derrien et al., 2014). Today, mycorrhizal fungi can associate with over 90% of higher plant species (including agricultural crops) and are found in soil of all continents from alpine lands to tropical forests and from grasslands to croplands.

AM symbiosis can promote plant growth and enhance plant resistance to a variety of abiotic stresses. Increased plant fitness and protection by AM symbiosis involve direct beneficial effects of AMF on structuring the rhizosphere, enhancing water and mineral nutrition while excluding toxic ions. AM symbiosis also changes plant gene expression and phytohormone balance affecting physiological processes related to plant metabolism, growth, and development (Panneerselvam et al., 2017).

Beneficial mycorrhizal effects on rice growth and resistance to abiotic stresses have often been observed in laboratory. Whether or not these effects can be repeated in much less controlled conditions, i.e., in field, and exploited by scientists to improve crop productivity remain unclear. Several reports have demonstrated that rice inoculation with AMF can improve plant growth and yield in field conditions.

8.3 CONCLUSION

Abiotic stresses hamper plant growth and productivity. Climate change and agricultural malpractices like excessive use of fertilizers and pesticides have aggravated the effects of abiotic stresses on crop productivity and degraded the ecosystem. There is an urgent need for environment-friendly management techniques such as the use of AMF for enhancing crop productivity and penetrating the cortical cells of the roots of a vascular plant forming arbuscules. AMF are soil microorganisms able to form mutualistic symbiosis with most terrestrial plants. Drought stress is responsible for the deterioration in the soil and presents severe threats to agriculture worldwide. Studies have confirmed that AM symbiosis-specific downstream responses control a combination of morphological, biochemical, and physiological plant characteristics. Although this can promote plant growth and enhance plant resistance to a variety of abiotic stresses, reports have demonstrated that rice inoculation with AMF can improve plant growth as well as yield in field conditions.

REFERENCES

Ahanger, M.A. and Agarwal, R.M. (2017). Salinity stress induced alterations in antioxidant metabolism and nitrogen assimilation in wheat (Triticum aestivum L) as influenced by potassium supplementation. *Plant Physiology and Biochemistry, 115*, 449–460.

Ahanger, M.A., Akram, N.A., Ashraf, M., Alyemeni, M.N., Wijaya, L. and Ahmad, P. (2017). Plant responses to environmental stresses—from gene to biotechnology. *AoB Plants, 9*(4).

Ahanger, M.A., Ashraf, M., Bajguz, A. and Ahmad, P. (2018). Brassinosteroids regulate growth in plants under stressful environments and crosstalk with other potential phytohormones. *Journal of Plant Growth Regulation, 37*, 1007–1024.

Begum, N., Qin, C., Ahanger, M.A., Raza, S., Khan, M.I., Ashraf, M.,... and Zhang, L. (2019). Role of arbuscular mycorrhizal fungi in plant growth regulation: implications in abiotic stress tolerance. *Frontiers in Plant Science, 10*, 1068.

Bonfante, P. and Genre, A. (2010). Mechanisms underlying beneficial plant–fungus interactions in mycorrhizal symbiosis. *Nature Communications, 1*(1), 1–11.

Garg, N. and Baher, N. (2013). Role of arbuscular mycorrhizal symbiosis in proline biosynthesis and metabolism of *Cicer arietinum* L. (chickpea) genotypes under salt stress. *Journal of Plant Growth Regulation, 32*(4), 767–778.

Herbinger, K., Tausz, M., Wonisch, A., Soja, G., Sorger, A. and Grill, D. (2002). Complex interactive effects of drought and ozone stress on the antioxidant defence systems of two wheat cultivars. *Plant Physiology and Biochemistry, 40*(6–8), 691–696.

Mitra, D., Be, G.S., Khoshru, B., De Los Santos Villalobos, S., Belz, C., Chaudhary, P., Shahri, F.N., Djebaili, R., Adeyemi, N.O., El-Ballat, E.M. and El-Esawi, M.A. (2021c). Impacts of arbuscular mycorrhizal fungi on rice growth, development, and stress management with a particular emphasis on strigolactone effects on root development. *Communications in Soil Science and Plant Analysis, 52*(14), 1591–1621.

Mitra, D., Chaudhary, P., Verma, D., Khoshru, B., Senapati, A., Mahakur, B., Panneerselvam, P., Mohapatra, P.K.D. and Anđelković, S. (2021a). Bioinformatics' role in studying microbe-mediated biotic and abiotic stress tolerance. In: A. Kumar and S. Droby (Eds.), *Microbial Management of Plant Stresses*. Woodhead Publishing, Sawston, pp. 203–219.

Mitra, D., Djebaili, R., Pellegrini, M., Mahakur, B., Sarker, A., Chaudhary, P., Khoshru, B., Gallo, M.D., Kitouni, M., Barik, D.P. and Panneerselvam, P. (2021b). Arbuscular mycorrhizal symbiosis: plant growth improvement and induction of resistance under stressful conditions. *Journal of Plant Nutrition, 44*(13), 1993–2028.

Mitra, D., Saritha, B., Janeeshma, E., Gusain, P., Khoshru, B., Nouh, F.A.A., Rani, A., Olatunbosun, A.N., Ruparelia, J., Rabari, A. and Mosquera-Sánchez, L.P. (2022). Arbuscular mycorrhizal fungal association boosted the arsenic resistance in crops with special responsiveness to rice plant. *Environmental and Experimental Botany, 193*, 104681.

Mitra, D., Uniyal, N., Panneerselvam, P., Senapati, A. and Ganeshamurthy, A.N. (2020). Role of mycorrhiza and its associated bacteria on plant growth promotion and nutrient management in sustainable agriculture. *International Journal of Life Sciences and Applied Sciences*, *1*(1), 1–1.

Panneerselvam, P., Kumar, U., Senapati, A., Parameswaran, C., Anandan, A., Kumar, A., Jahan, A., Padhy, S.R. and Nayak, A.K. (2020). Influence of elevated CO_2 on arbuscular mycorrhizal fungal community elucidated using Illumina MiSeq platform in sub-humid tropical paddy soil. *Applied Soil Ecology*, *145*, 103344.

Panneerselvam, P., Kumar, U., Sugitha, T.C.K., Parameswaran, C., Sahoo, S., Binodh, A.K., Jahan, A. and Anandan, A. (2017) Arbuscular mycorrhizal fungi (AMF) for sustainable rice production. In: T.K. Adhya, B. Lal, B. Mohapatra, D. Paul and S. Das (Eds.), *Advances in Soil Microbiology: Recent Trends and Future Prospects*. Springer, Singapore, pp. 99–126.

Panneerselvam, P., Mohandas, S., Saritha, B., Upreti, K.K., Poovarasan, Monnappa, A. and Sulladmath, V.V. (2012) Glomus mosseae associated bacteria and their influence on stimulation of mycorrhizal colonization, sporulation, and growth promotion in guava (*Psidium guajava* L.) seedlings. *Biological Agriculture & Horticulture*, *28*(4), 267–279.

Panneerselvam, P., Sahoo, S., Senapati, A., Kumar, U., Mitra, D., Parameswaran, C., Anandan, A., Kumar, A., Jahan, A. and Nayak, A.K. (2019). Understanding interaction effect of arbuscular mycorrhizal fungi in rice under elevated carbon dioxide conditions. *Journal of Basic Microbiology, 59*(12), 1217–1228.

Panneerselvam, P., Saritha, B., Mohandas, S., Upreti, K.K., Poovarasan, S., Sulladmath, V.V. and Venugopalan, R. (2013). Effect of mycorrhiza-associated bacteria on enhancing colonization and sporulation of Glomus mosseae and growth promotion in sapota (Manilkara achras (Mill) Forsberg) seedlings. *Biological Agriculture & Horticulture*, *29*(2), 118–131.

Remy, W., Taylor, T.N., Hass, H. and Kerp, H. (1994). Four hundred-million-year-old vesicular arbuscular mycorrhizae. *Proceedings of the National Academy of Sciences*, *91*(25), 11841–11843.

Santander, C., Sanhueza, M., Olave, J., Borie, F., Valentine, A. and Cornejo, P. (2019). Arbuscular mycorrhizal colonization promotes the tolerance to salt stress in lettuce plants through an efficient modification of ionic balance. *Journal of Soil Science and Plant Nutrition*, *19*(2), 321–331.

Saritha, B., Panneerselvam, P., Srinivas, K., Debasis, M. and Ansuman, S. (2021). Enhancing the sapota [Manilkara achras (Mill) Forsberg] yield through intervention of Arbuscular mycorrhizal fungi and its associated bacteria. *Research Journal of Biotechnology*, *16*, 99–109.

Smith, S.E. and Read, D.J. (2008). *Mycorrhizal Symbiosis*. 3rd Edition, Academic Press, London.

Strullu-Derrien, C., Kenrick, P., Pressel, S., Duckett, J.G., Rioult, J.P. and Strullu, D.G. (2014). Fungal associations in H orneophyton ligneri from the R hynie C hert (c. 407 million year old) closely resemble those in extant lower land plants: novel insights into ancestral plant–fungus symbioses. *New Phytologist*, *203*(3), 964–979.

Stürmer, S.L. (2012). A history of the taxonomy and systematics of arbuscular mycorrhizal fungi belonging to the phylum Glomeromycota. *Mycorrhiza, 22*(4), 247–258.

Taylor, T.N., Remy, W., Hass, H. and Kerp, H. (1995). Fossil arbuscular mycorrhizae from the Early Devonian. *Mycologia*, *87*(4), 560–573.

Zhao, R., Guo, W., Bi, N., Guo, J., Wang, L., Zhao, J. and Zhang, J. (2015). Arbuscular mycorrhizal fungi affect the growth, nutrient uptake and water status of maize (*Zea mays* L.) grown in two types of coal mine spoils under drought stress. *Applied Soil Ecology*, *88*, 41–49.

9 Arbuscular Mycorrhiza and Its Role in Rice Production under Drought Stress

Biswajit Pramanik and Puranjoy Sar
Visva-Bharati

Shreosi Biswas and Partha Chandra Mondal
Indian Agricultural Research Institute

CONTENTS

9.1 INTRODUCTION

Rice (*Oryza sativa* L.) is one of the highest produced cereals, which is consumed by half of the world's population. Specifically, in the South Asian nations, it is grown as a prime food grain. Notably, the global population is about to cross 10 billion in 2050, and to secure food for each individual of this over-expanding population, the production of food crops as well as rice should be doubled compared to the present. Alongside its wide spread, this crop also suffers from various biotic and abiotic stresses nowadays. Drought, caused by a deficiency of water, becomes one of the major adverse abiotic factors affecting the productivity of rice, especially in rainfed areas (Aslam et al., 2022). Changes in climatic conditions due to human interference have made this stress more severe in semi-arid and arid zones. Apart

DOI: 10.1201/9781003354086-9

from disturbing osmotic balance, membrane components distribution, etc., drought stress also triggers reactive oxygen species (ROS) generation (Zou et al., 2021). The elevation in ROS concentration, along with downregulation of various non-cyclic processes cumulatively, brings about oxidative damage to the plant and obstruction to ATP synthesis, respectively. This eventually intensifies the rate of photorespiration, which ultimately leads to declining photosynthesis that finally affects the yield (Ullah et al., 2017). Therefore, to combat these stress conditions, various alternative strategies can be adapted such as developing drought-tolerant or genome-modified varieties or applying various symbiotic microorganisms. Among them, arbuscular mycorrhizal fungi (AMF) occupy the first row in terms of forming a symbiotic mycorrhizal association with most of the terrestrial plants and aid several physiological and biochemical processes, i.e., photosynthesis, root development, stress resistance, nutrient uptake, and many more with higher proficiency than others (Bernaola and Stout, 2019). Hence, in a nutshell, this chapter predominantly focuses on AMF, their key specifications, and potency with the mechanisms followed by them to withstand water-deficit conditions, with special attention to rice.

9.2 AMF: OVERVIEW WITH HIERARCHICAL POSITION

According to the taxonomical classification, AMF reside under the phylum "Glomeromycota." This phylum possesses a single class "Glomeromycetes" containing four orders, namely, "Glomerales," "Archeosporales," "Diversisporales," and "Paraglomerales." The commonly found genera within AMF are *Glomus*, *Scutellospora*, *Gigaspora*, and *Entrophospora*. These fungi show obligate symbiosis in the roots of the host plants. Although host-specific AMF are not found till date, host preferences were conferred by them (Bagyaraj, 2014; Panneerselvam et al., 2019, 2020). Some kind of distinctiveness has been detected in the symbiont formation process of AMF as some portions of it (vesicles, arbuscles, and intraradical hyphae) reside inside the root, and the rest (extra-matricular chlamydospores and extraradical hyphae) is found within the soil. Their increased population of AMF in the soil also helps in improving fertility, providing resistance to soilborne pathogens, and enhancing the acquisition of diffusion-controlled nutrients, i.e., Zn, P, Cu, etc. (Bagyaraj, 2014; Mitra et al., 2021a; Panneerselvam et al., 2017). Most of the terrestrial crops prevail in a mutualistic relationship with these root symbionts to secure themselves in terms of nutrient supply and stress resistance in exchange of providing shelter. Both partners in this symbiosis are involved in maintaining ecological balance (Mathimaran et al., 2017). The responses to alleviated water-deficit stress conditions by plants with the help of AMF reportedly have an accumulative impact on the morphological, physiological, and cytological mechanisms (Ruiz-Lozano, 2003).

9.3 AMF ON DROUGHT TOLERANCE IN RICE

Drought stress induces enormous adverse effects at all stages of rice cultivation. Some tolerant rice genotypes trigger several complex response mechanisms

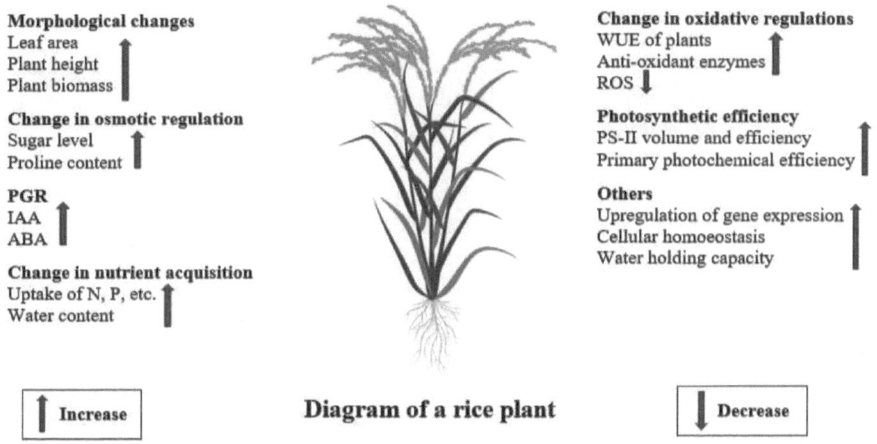

FIGURE 9.1 Effect and mechanism followed by rice plant due to the presence of the symbiotic association with AMF under drought stress. (*Source: Self-created.*)

during fundamental processes to combat moisture stress (Figure 9.1). These include morphology-oriented responses, such as drought escape and avoidance, and physiological events such as the accumulation of various osmolytes and antioxidants, and phytohormonal homeostasis (Jangra et al., 2019). AMF, endophytic fungi, colonize in the fibrous root system of rice to enhance the plant's tolerance response mechanisms while undergoing moisture stress. The following section will depict the effect of the association between AMF and rice, and its advantages under drought stress conditions.

9.3.1 Plant-Water Retention

The maintenance of water status inside a biological system is a necessity for proper continuation of various metabolic processes in a stressed environment. Tolerant cultivars can acquire a wide array of strategies for maintaining higher turgor pressure to withstand the same. Therefore, maintaining intracellular water status has a lot of relevance in maintaining various physiological processes and in improving the growth and productivity of plants in stress environment. Following this direction, AMF also offer protection to the plants by assisting plants to sustain optimum turgor pressure under moisture stress. Numerous experimental studies revealed that the relative water content and water potential of the plants are significantly improved by AMF. Apart from maintaining an optimal cytological aqueous condition, this mycorrhizal association also provides an advantage to the crop plants by enhancing water use efficiency (WUE) (Zhu et al., 2012). This symbiosis improves the cellular aquatic status in plants leading to enhanced water and nutrient uptake, which improves root architecture and enhanced rhizosphere activity (Sheng et al., 2008). Consequently, plants become physically and physiologically sturdy to cope up with the stressed situation.

9.3.2 Osmotic Balance

One of the key strategies adopted by plants to combat against drought is osmotic adjustment. This helps to maintain the turgor pressure inside plants. While undergoing such stress conditions, plants usually generate several osmolytes *viz.* glycine-betaine, proline, and sugar alcohol derivatives. Under water deficiency, plants significantly enhance the production of these osmolytes.

In several crop species, this plant-AMF association also succored the accumulation of many osmolytes upon stress, as reported. These osmolytes, under such stress situations, play a key role not only by adjusting the intracellular osmotic environment but also by serving as a potent carbon (C) source. This shows the significance of plant-AMF association, which triggers various pathways in plants that are responsible for osmotic adjustment to balance the plant-water potential. Moreover, it is already well established that the genotypes with the ability to produce higher amounts of proline can tolerate the water-deficit condition efficiently. Augé (2001) reported the commendable accumulation of proline in AM-inoculated plants. Because of the elevated osmotic regulation, the turgor pressure was detected to be greater in AMF-associated plants rather than the non-AMF-associated plants. Hence, various turgor-dependent processes such as cellular growth and expansion, opening of guard cells surrounding stomatal pores, and photosynthesis coupled with balanced water potential gradient for effective permeation of capillary water into crops through the root system can be ascertained (Ruiz-Lozano, 2003). Ocón et al. (2007) reported that significant trehalose accumulation occurred during plant-AMF associations. Trehalose, a versatile non-reducing disaccharide, plays a pivotal role in stabilizing cell structures and acts as a long-lived chaperon. Under drought stress, they help conserve the native conformation of protein, resulting in greater membrane stability, which leads to reduced leakage of electrolytes.

9.3.3 Generation of Antioxidants

When a plant experiences moisture stress, ROS are produced in a profuse amount, resulting in oxidative damage to the plant itself. ROS are free radicals that lead to electrolyte leakage and oxidative erosion to the lipids, proteins, and DNA both inside and outside the cell (Zou et al., 2021). Upon mismanaging, these molecules may cause irreversible damage to the plasma membranes, which may eventually inhibit various cellular metabolic pathways. Hence, effective ROS quenching is very critical to maintain cellular metabolism and survival under harsh situations. Stress-resistant plants possess the ability to manage ROS through increased antioxidant system activity. Decreased ROS generation and oxidative injury in AM-associated crops are the consequences of enhanced activity of various enzymes, namely, superoxide dismutase (SOD), peroxidase, and catalase, which were found to be in higher concentrations in those plants in such environment (Khalafallah and Abo-Ghalia, 2008). The significant quenching of ROS by antioxidant enzymes reduced its level in AM-associated plants. Further to mention, both the antioxidative enzyme and non-enzymatic systems are responsible for the depletion in ROS concentration in those plants. Enhanced stability in cell membranes due to the changes in membrane phospholipids levels, reduced lipid peroxidation, and H_2O_2 accumulation under moisture stress were also reported by Chandrasekaran and Paramasivan (2022).

9.3.4 Photosynthetic Behavior

Photosynthesis is primarily affected by plant's exposure to stressed conditions. Following the minimal photosynthetic efficiency along with leaf area shrinkage, the net available carbon decreases lesser, which causes significant yield reduction in stress-susceptible plants. Ruiz-Sánchez et al. (2010) discovered that *Rhizophagus intraradices*-inoculated rice exhibited higher photosynthetic efficacy during drought situations. Zhu et al. (2012) observed enhanced transpiration as well as photosynthetic rates in crops inoculated with AMF. The stomatal conductance in such plants provides a wider opening of the stomata to facilitate the gaseous exchange for transpiration and photosynthesis (Tyagi et al., 2017). As the AMF colonized plants own the capability to absorb capillary water for supplying in the transpiration pull, the rate of transpiration gets increased in those crops (Roumet et al., 2006). The chlorophyll content of the host plant also contributes to the increased rate of photosynthesis due to this symbiotic fungal relationship. Alongside augmenting the chlorophyll content, the mutualistic association also protects the pigment from being degraded under the mentioned abiotic stress (Chareesri et al., 2020). In brief, the beneficial act of protecting the host plants from water-scarce conditions due to the presence of this symbiosis is briefly described in this section.

9.4 MECHANISMS OF DROUGHT TOLERANCE

The mutualistic relationship between plant roots and AMF not only strengthens soil nutritional factors but also helps plants adapt more efficiently to osmotic stress conditions. This adaptation may follow several mechanisms (Figure 9.1). Osmotic stress triggers the production of aquaporins (AQPs), which act as water channels and, thus, are considered one of the most important proteins in osmoregulation. AQPs, in accordance with the water potential gradient, accelerate the passive transport pathways of water. In addition to the universal solvent, these proteins also transport several smaller ions and molecules *viz.* ammonium (NH_4^+), glycerol, and carbon dioxide (CO_2). Thus, positive gene regulation of AQP-PIPs (aquaporin-plasma membrane intrinsic proteins) in fungi-colonized plants can multiply WUE up to several folds by reducing ROS cumulation and diminishing oxidative degeneration.

Under stress condition, the microbes-associated symbiosis improves the resistance capability of the host plants against stresses and, thus, helps them overcome the same (Bilal et al., 2020). Generally, the plant-water retention capacity of the host plant is induced by the association with AMF. The methodologies include change in hormonal and osmotic regulations, effective soil-root contact, increased soil water absorption and gaseous exchange efficiency, mycelium-mediated direct uptake of water from soil and its transport to the sink, etc. (Augé, 2001). Physically, the interaction between extraradical mycelia and the soil particles helps in aggregation to form a stable compound to retain the water with the help of soil organic matter (Mardhiah et al., 2016).

While exploring the biochemical perspective, it has been demonstrated that AMF are able to release glomalin or glomalin-related soil protein (GRSP), several polysaccharides, mucilage, hydrophobins, etc., in soil (Singh et al., 2012). Also, production of abscisic acid (ABA) in higher concentration, due to the upregulation of ABA signaling pathway, causes upliftment of hydraulic conductivity to balance plant-water relation. Thus, both the physical and biochemical effects of AMF aid in the generation of stable soil aggregates, which eventually improve water retention along with increased infiltration capacity of the respective soil, resulting in drought retardation in the rhizosphere of the host plant (Manoharan et al., 2010; Audet, 2012). AM mycelia penetrate the soil pores to absorb and transport water, unavailable to plant roots, from soil to the plant body. Further to indicate, phosphorus (P) acquisition has been improved by the formation of plant-AMF association. Recent evidence has proposed that AMF-AQPs actively participate in the polyphosphate translocation process via fungal hyphae. The application of AM assisted in regaining the mobility of nitrate (NO_3^-) and NH_4^+, along with assimilating the immobile element P, under osmotic stress. The increased nitrogen (N) percentage inside the plant body promotes protein biosynthesis and accumulation of compatible osmolytes and, therefore, helps in boosting plant metabolism pathways. Notably, these fungi also expand the surface area to volume ratio of roots for superior absorption and nutrient acquisition capacity and, thus, promote enhanced plant growth (Mitra et al., 2021b).

Physiologically, intense AMF-associated rice plant exhibits higher accumulation of biomass, which ultimately contributes to the overall performance of the plant. This interaction is also involved in improving various bio-physiological processes, i.e., increasing photosynthetic efficiency by elevating the volume of PS-II as well as improvement in the architecture or hydraulic conductivity of roots (Jajoo and Mathur, 2021). Better photosynthetic ability was also achieved through the accumulation of antioxidants such as glutathione, due to AMF association. Several studies have shown that antioxidant genes (such as GintMT1, GintPDX1, and GintSOD1) from AMF participate in mitigating ROS accumulation. ROS scavenging genes, including glutathione peroxidase, were specifically related to the detoxification process of the same. Basically, glutathione reduces the concentration of H_2O_2 and lipids inside the cytosol and on the cell membrane, respectively (Ruiz-Sánchez et al., 2010). Rather, wider conductance of stomatal pores for better evaporation to create a transpiration pull inside the xylem tissue is mediated by mutualism. This transpiration pull helps in uptaking more capillary water through the roots. Succinctly, AMF-rice association reportedly helps the host plant to resist the above-mentioned abiotic stress condition by improvising several aspects.

9.5 CONCLUSION AND FUTURES

Each environmental stress, including both abiotic and biotic, causes a greater reduction in productivity by altering numerous physiological and biochemical processes related to plant metabolism. Plants generally use several strategies not only to survive these conditions but also to generate higher yields that overcome these adverse situations. Drought or moisture stress is one of them, which is the most important abiotic stress under such rapidly changing climatic conditions. Similar to other crops, rice, the most important cereal, also undergoes it and faces a huge amount of yield loss. AMF can efficiently alleviate the harmful effects caused by drought stress on rice. These factors help acclimatize the plant by providing better buffering capacity under stress condition. To date, many studies have been conducted to investigate the beneficial effects of the symbiotic association of AMF with rice. These fungi form a mycorrhizal association with most of the land-dwelling plants. From this relationship, the host plant gets benefits in terms of resisting the deficiency of water, pathogen, and/or insect infestation, while the fungal portion is provided shelter. Further research needs to be executed to identify the particular AMF species, which provide the optimum symbiosis with rice under moisture stress. This chapter predominantly focused on the rewarding consequences of this cooperative association with special references to the physical, physiological, biochemical, and molecular mechanistic approaches. Although very little knowledge has been perceived through past research, digging up at a profoundly minute level is remaining. Therefore, this aforesaid chapter may help the upcoming researchers by encouraging them to explore it to a much extensive extent for the sake of human welfare.

REFERENCES

Aslam, M.M., Rashid, M.A.R., Siddiqui, M.A., Khan, M.T., Farhat, F., Yasmeen, S., Khan, I.A., Raja, S., Rasool, F., Sial, M.A. and Yan, Z., 2022. Recent insights into signaling responses to cope drought stress in rice. *Rice Science 29*(2): 105–117. https://doi.org/10.1016/j.rsci.2021.08.001.

Audet, P., 2012. Arbuscular mycorrhizal symbiosis and other plant–soil interactions in relation to environmental stress. In *Environmental Adaptations and Stress Tolerance of Plants in the Era of Climate Change*, eds. P. Ahmad, and M.N.V. Prasad, pp. 233–264. Springer, New York. https://doi.org/10.1007/978-1-4614-0815-4_11.

Augé, R.M., 2001. Water relations, drought and vesicular-arbuscular mycorrhizal symbiosis. *Mycorrhiza 11*(1): 3–42. https://doi.org/10.1007/s005720100097.

Bagyaraj, D.J., 2014. Mycorrhizal fungi. *Proceedings of the Indian National Science Academy 80*(2): 415–428.

Bernaola, L. and Stout, M.J., 2019. Effects of arbuscular mycorrhizal fungi on rice-herbivore interactions are soil-dependent. *Scientific Reports 9*(1): 1–12. https://doi.org/10.1038/s41598-019-50354-2.

Bilal, S., Shahzad, R., Imran, M., Jan, R., Kim, K.M. and Lee, I.J., 2020. Synergistic association of endophytic fungi enhances *Glycine max* L. resilience to combined abiotic stresses: heavy metals, high temperature and drought stress. *Industrial Crops and Products 143*: 111931. https://doi.org/10.1016/j.indcrop.2019.111931.

Chandrasekaran, M. and Paramasivan, M., 2022. Arbuscular mycorrhizal fungi and antioxidant enzymes in ameliorating drought stress: a meta-analysis. *Plant and Soil 8*: 1–9. https://doi.org/10.3390/jof8070660.

Chareesri, A., De Deyn, G.B., Sergeeva, L., Polthanee, A. and Kuyper, T.W., 2020. Increased arbuscular mycorrhizal fungal colonization reduces yield loss of rice (*Oryza sativa* L.) under drought. *Mycorrhiza 30*(2): 315–328. https://doi.org/10.1007/s00572-020-00953-z.

Jajoo, A. and Mathur, S., 2021. Role of arbuscular mycorrhizal fungi as an underground saviuor for protecting plants from abiotic stresses. *Physiology and Molecular Biology of Plants* 27(11): 2589–2603. https://doi.org/10.1007/s12298-021-01091-2.

Jangra, S., Mishra, A., Disha, K., Yadav, N.R. and Yadav, R.C., 2019. Plant responses and tolerance to drought. In *Approaches for Enhancing Abiotic Stress Tolerance in Plants*, eds. M. Hasanuzzaman, K. Nahar, M. Fujita, H. Oku and T. Islam, pp. 79–98. CRC Press, Boca Raton, FL.

Khalafallah, A.A. and Abo-Ghalia, H.H., 2008. Effect of arbuscular mycorrhizal fungi on the metabolic products and activity of antioxidant system in wheat plants subjected to short-term water stress, followed by recovery at different growth stages. *Journal of Applied Sciences Research 4*(5): 559–569.

Manoharan, P.T., Shanmugaiah, V., Balasubramanian, N., Gomathinayagam, S., Sharma, M.P. and Muthuchelian, K., 2010. Influence of AM fungi on the growth and physiological status of *Erythrina variegata* Linn. grown under different water stress conditions. *European Journal of Soil Biology 46*(2): 151–156. https://doi.org/10.1016/j.ejsobi.2010.01.001.

Mardhiah, U., Caruso, T., Gurnell, A. and Rillig, M.C., 2016. Arbuscular mycorrhizal fungal hyphae reduce soil erosion by surface water flow in a greenhouse experiment. *Applied Soil Ecology 99*: 137–140. https://doi.org/10.1016/j.apsoil.2015.11.027.

Mathimaran, N., Sharma, M.P., Mohan Raju, B. and Bagyaraj, D.J., 2017. Arbuscular mycorrhizal symbiosis and drought tolerance in crop plants. *Mycosphere 8*(3): 361–376. https://doi.org/10.5943/mycosphere/8/3/2.

Mitra, D., Be, G.S., Khoshru, B., De Los Santos Villalobos, S., Belz, C., Chaudhary, P., Shahri, F.N., Djebaili, R., Adeyemi, N.O., El-Ballat, E.M. and El-Esawi, M.A., 2021a. Impacts of arbuscular mycorrhizal fungi on rice growth, development, and stress management with a particular emphasis on strigolactone effects on root development. *Communications in Soil Science and Plant Analysis 52*(14): 1591–1621. https://doi.org/10.1080/00103624.2021.1892728.

Mitra, D., Djebaili, R., Pellegrini, M., Mahakur, B., Sarker, A., Chaudhary, P., Khoshru, B., Gallo, M.D., Kitouni, M., Barik, D.P. and Panneerselvam, P., 2021b. Arbuscular mycorrhizal symbiosis: plant growth improvement and induction of resistance under stressful conditions. *Journal of Plant Nutrition 44*(13): 1993–2028.

Ocón, A., Hampp, R. and Requena, N., 2007. Trehalose turnover during abiotic stress in arbuscular mycorrhizal fungi. *New Phytologist 174*(4): 879–891. https://doi.org/10.1111/j.1469-8137.2007.02048.x.

Panneerselvam, P., Kumar, U., Senapati, A., Parameswaran, C., Anandan, A., Kumar, A., Jahan, A., Padhy, S.R. and Nayak, A.K., 2020. Influence of elevated CO_2 on arbuscular mycorrhizal fungal community elucidated using Illumina MiSeq platform in sub-humid tropical paddy soil. *Applied Soil Ecology 145*: 103344.

Panneerselvam, P., Kumar, U., Sugitha, T.C.K., Parameswaran, C., Sahoo, S., Binodh, A.K., Jahan, A. and Anandan, A., 2017. Arbuscular mycorrhizal fungi (AMF) for sustainable rice production. In *Advances in Soil Microbiology: Recent Trends and Future Prospects*, eds. T.K. Adhya, B. Lal, B. Mohapatra, D. Paul, and S. Das, pp. 99–126. Springer, Singapore.

Panneerselvam, P., Sahoo, S., Senapati, A., Kumar, U., Mitra, D., Parameswaran, C., Anandan, A., Kumar, A., Jahan, A. and Nayak, A.K., 2019. Understanding interaction effect of arbuscular mycorrhizal fungi in rice under elevated carbon dioxide conditions. *Journal of Basic Microbiology 59*(12): 1217–1228.

Roumet, C., Urcelay, C. and Díaz, S., 2006. Suites of root traits differ between annual and perennial species growing in the field. *New Phytologist 170*(2): 357–368. https://doi.org/10.1111/j.1469-8137.2006.01667.x.

Ruiz-Lozano, J.M., 2003. Arbuscular mycorrhizal symbiosis and alleviation of osmotic stress. New perspectives for molecular studies. *Mycorrhiza 13*(6): 309–317. https://doi.org/10.1007/s00572-003-0237-6.

Ruiz-Sánchez, M., Aroca, R., Muñoz, Y., Polón, R. and Ruiz-Lozano, J.M., 2010. The arbuscular mycorrhizal symbiosis enhances the photosynthetic efficiency and the antioxidative response of rice plants subjected to drought stress. *Journal of Plant Physiology 167*(11): 862–869. https://doi.org/10.1016/j.jplph.2010.01.018.

Sheng, M., Tang, M., Chen, H., Yang, B., Zhang, F. and Huang, Y., 2008. Influence of arbuscular mycorrhizae on photosynthesis and water status of maize plants under salt stress. *Mycorrhiza 18*(6): 287–296. https://doi.org/10.1007/s00572-008-0180-7.

Singh, S., Gupta, A.K. and Kaur, N., 2012. Differential responses of antioxidative defence system to long-term field drought in wheat (*Triticum aestivum* L.) genotypes differing in drought tolerance. *Journal of Agronomy and Crop Science 198*(3): 185–195. https://doi.org/10.1111/j.1439-037X.2011.00497.x.

Tyagi, J., Sultan, E., Mishra, A., Kumari, M. and Pudake, R.N., 2017. The impact of AMF symbiosis in alleviating drought tolerance in field crops. In *Mycorrhiza-Nutrient Uptake, Biocontrol, Ecorestoration*, eds. A. Varma, R. Prasad and N. Tuteja, pp. 211–234. Springer, Cham. https://doi.org/10.1007/978-3-319-68867-1_11.

Ullah, A., Sun, H., Yang, X. and Zhang, X., 2017. Drought coping strategies in cotton: increased crop per drop. *Plant Biotechnology Journal* 15(3): 271–284. https://doi.org/10.1111/pbi.12688.

Zhu, X.C., Song, F.B., Liu, S.Q., Liu, T.D. and Zhou, X., 2012. Arbuscular mycorrhizae improves photosynthesis and water status of *Zea mays* L. under drought stress. *Plant, Soil and Environment 58*(4): 186–191. https://doi.org/10.17221/23/2011-PSE.

Zou, Y.N., Wu, Q.S. and Kuča, K., 2021. Unravelling the role of arbuscular mycorrhizal fungi in mitigating the oxidative burst of plants under drought stress. *Plant Biology 23*: 50–57. https://doi.org/10.1111/plb.13161.

10 Arbuscular Mycorrhiza and Its Role in Rice Production under Salinity Stress

Shampa Purkaystha
Centurion University of Technology and Management

Biswajit Pramanik
Visva-Bharati

Anamika Das
Bidhan Chandra Krishi Viswavidyalaya

CONTENTS

10.1 INTRODUCTION

The field-grown plants are often introduced to a combination of several environmental stress conditions *viz.* drought, salinity, cold, heat, high temperature, flood, UV exposure, heavy metals, etc. (Mitra et al., 2021a). Among all, salinity predominantly plays a major role in affecting the growth and development of cultivated crops. When crops are subjected to salinity stress, they exhibit ionic imbalance, retarded seedling germination, and growth, which ultimately leads to poor seed set at the developmental stage (Kumar et al., 2010; Mitra et al., 2021b). Rice (*Oryza sativa* L.), the most important monocarpic cereal across the world, especially in the South Asian nations, is also not devoid of the adverse effects of this particular abiotic stress (Panneerselvam et al., 2019). Almost 40 crore tons of rice are produced annually to serve half of the global

DOI: 10.1201/9781003354086-10

population. Thus, this crop not only improves livelihoods but also alleviates poverty throughout Asia and the entire world (Muthuramalingam et al., 2022). Recently, salt stress has evolved as one of the most devastating constraints during the growth stages of rice, specifically at the seedling and blossoming stages (Hosseini et al., 2012), apart from affecting the yield-attributed quantitative traits such as the number of tillers, panicles and spikelet, weight of a single seed, seeds per panicle as well as an entire plant, harvest index, and fertility percentage. Salinity intensely hinders different physiological processes like photosynthesis, evapotranspiration, and stomatal movement. The accumulation of salt at much higher concentrations is also responsible for the disruption of chlorophyll, nitrogen (N), and phosphorus (P) acquisition at a lower rate, and accumulation of lesser biomass (Jan et al., 2018; Mitra et al., 2021c). At the biochemical level, overproduction of reactive oxygen species (ROS) not only induces oxidative damage to the cells but also transfers stress signals by acting as secondary messengers (Khaliq et al., 2015). To combat such negative impacts of higher salinity concentrations in the soil as well as inside plant body, several strategies have been adapted following numerous mechanisms including generation of antioxidants, enhancement of nutrient uptake efficiency, etc. Reviewing all the recent studies and considering the aforesaid strategies, it was found that arbuscular mycorrhizal fungi (AMF) may be deployed as an efficient, sustainable, inexpensive, and eco-friendly approach toward combating salinity stress in rice cultivation (Panneerselvam et al., 2019, 2020). Hence, the following chapter discusses the usage and mechanistic approaches of AMF in withstanding the adverse effects of salinity in rice.

10.2 AMF AND THEIR ASSOCIATION WITH RICE

AMF belong to the phylum Glomeromycota, and they own a special niche in the soil microbiome. They set up a mutualistic relationship in the roots of the most terrestrial plants, involving many agricultural food crops, although preferences for particular hosts are reported (Bagyaraj, 2014). The AMF species showing potent symbiosis with rice include *Glomus geosporum*, *Glomus mosseae*, *Glomus intraradices*, *Scutellospora* sp., and *Acaulospora* sp. As rice provides food security to an enormous population, its cultivation should be free from various pathogens and insect infestation. The disease and pest management are mediated by the usage of numerous chemicals, which is creating environmental safety concerns and the generation of super-pathogens and super-pests, nowadays. Efficient uptake of macro- and micronutrients, especially carbon, nitrogen, phosphorous, potassium, zinc, and boron, can equip the crop with the ability to resist pathogen attack (Panneerselvam et al., 2017). On the background of these sequences of events, the application of AMF has been proven to be a significant step to offer optimum growth, development, and overall yield of rice plant by not only acquiring nutrients from the soil but also improving soil structure and salinity stress. Therefore, the following sections will elaborately depict the effects and mechanisms of AMF colonization to check the harmful effects of salt cumulation on rice.

10.3 AMF AND SALINITY STRESS IN RICE

Salinity stress is the major problem, mostly in the coastal areas, which limits the growth of the crop and ultimately hampers productivity. Also, the presence of glycophyte in a rice plant makes it more susceptible to excess saline condition, especially in the early vegetative and late reproductive stages (Ghosh et al., 2016). Among all other biological interventions applied to reduce the negative impact of salt stress on this cereal, the employment of AMF is considered the most efficient. These microorganisms act through various biochemical and physiological mechanisms. Therefore, plant-AMF association not only helps reduce the negative effect of this specific abiotic stress but also increases the plant biomass, root growth, and eventually the overall growth of the plant. During stress situations, AMF increase the integrity of the cell wall, which is a potent reason behind optimum cellular development (Qin et al., 2021). Apart from these, from a physiological perspective, this symbiosis provides the plant with better photosynthetic efficiency along with escalated osmotic adjustable capacity, which, eventually, balances the plant-water retention ability.

10.4 AMF-ASSISTED MECHANISMS TO OVERCOME SALINITY STRESS

Depending on the cultivar, salinity significantly decreases the uptake of phosphorus and imbalances the Na^+/K^+ pump protein (Su et al., 2022). Phosphorus is one of the essential elements that affect the growth and development of plant roots, while K^+ accumulation in the root zone during salinity stress gives it enough strength to combat various biotic stressors too. Phosphorus, which is expected to be one of the most limiting elements by 2050, is absorbed directly by the plant roots in the form of orthophosphate (PO_4^{3-}). This compound remains bound to the soil particles during adverse conditions creating phosphorus limitation, while in the presence of AMF, plants can absorb this element in the form of inorganic phosphate. The enzyme phosphatase and metabolites such as siderophores, released from rhizosphere, both help solubilize the soil phosphorus in the available form from poorly available phosphorus source like rock phosphate (Shenoy and Kalagudi, 2005) and change the soil pH too (Li and Christie, 2001). The availability of phosphorus mainly depends on the surface area of the interacting AMF-associated plant, which indirectly affects the symbiosis of AMF by allocating the less available carbon from the host plant to the fungus (Johnson, 2010). AMF are aerobic, and thus, they help the rice plant to develop aerenchyma cells during waterlogged conditions (Wang et al., 2013). Colonization depends on the growth stages of the rice as during the heading and ripening stages, the growth of vesicular (storage organ) and arbuscular (site where exchange of material occurs) structures is significantly more due to the development of the aerenchyma cells in the roots. Hence, in the conventional rice cultivation system, the application of AMF is important because of their positive nature regarding the aerenchyma cells' development, wide agrobiodiversity, and their adaptability. Among all

AMF, *Glomus* is the leading species both in soil and roots (Wang et al., 2015). Different AMF increase various plant growth-regulating substances *viz.* IAA, cytokinins, gibberellins, Vitamin B, etc. The AMF-plant relationship also alters the plant-water relation due to which plants can grow in osmotic stress conditions. This is because AMF possess very fine hypha (diameter of 2–5 nm) compared to the root hairs (diameter of 10–20 nm), which smoothen the water transport route within the plant body (Allen, 1982). Moreover, these hyphal structures prevent dehydration by enhancing the solute potential, which, at the end, not only reduces the osmotic potential but also increases the tolerance capacity in rice (Augé et al., 1987). This symbiosis is also important for the continuation of several metabolic processes in plants by improving different physiological and biochemical activities, such as elevated stomatal conductance, accumulation of proline, higher photosynthetic efficiency, and production of antioxidants (Faber et al., 1991). Further to mention, the inoculation of *Glomus etunicatum* and *Funneliformis geosporum* proves the host specificity of AMF, where the former one increases the content of chlorophyll a, proline, sugar, etc. and the latter is associated with phosphorus retention in leaves (Tisarum et al., 2020). Some AMF species (e.g., *Acaulospora mellea*, *Glomus formosanum*, *Rhizoglomus clarum*, and *Glomus* spp.) also help in the proper growth of the root morphology to resist salinity.

10.5 CONCLUSION AND FUTURE OUTLOOK

Salinity stands out to be a prevalent abiotic stress in terms of its daunting effects on rice production, especially during the seedling and flowering stages. To address salt stress in rice farming, AMF can be used as an effective, low-cost, and environmentally benign solution. In the plant roots, they create a mutualistic interaction by drawing nutrients from the soil and enhancing soil quality. Also, this symbiosis provides the plant with more osmotic adaptability, increased cell wall integrity, and improved photosynthetic efficiency to survive in stress conditions. Some enzymes, substances, and metabolites released by AMF in the root zone of rice help in better nutrient and water uptake and improve the phosphorus availability and the plant's defense mechanism. To date, only a few species of AMF have been studied in relation to rice production. Therefore, other species can be investigated for their role in rice production. There is huge scope for research on the role of AMF in upland and aerobic systems of rice production, determining the species specific to rice and important crops. Moreover, molecular markers could be developed for assessing the effectiveness of AMF in different growth stages of crops, which will smoothen the strategic pathways for future researchers.

AUTHORS' CONTRIBUTIONS

Conceptualization: SP, BP, and AD; Basic manuscript preparation: SP and BP; Addition and correction: AD; Final edit: BP and AD; Finalization: SP, BP, and AD. All authors read and approved the final manuscript.

REFERENCES

Allen, M.F. 1982. Influence of vesicular-arbuscular mycorrhizae on water movement through *Bouteloua gracilis* (HBK) Lag ex Steud. *New Phytologist 91*(2): 191–96. https://doi.org/10.1111/j.1469-8137.1982.tb03305.x.

Augé, R.M., Schekel, K.A. and Wample, R.L. 1987. Leaf water and carbohydrate status of VA mycorrhizal rose exposed to drought stress. *Plant and Soil 99*(2): 291–302. https://doi.org/10.1007/BF02370876.

Bagyaraj, D.J. 2014. Mycorrhizal fungi. *Proceedings of the Indian National Science Academy 80*(2): 415–28.

Faber, B.A., Zasoski, R.J., Munns, D.N. and Shackel, K. 1991. A method for measuring hyphal nutrient and water uptake in mycorrhizal plants. *Canadian Journal of Botany 69*(1): 87–94. https://doi.org/10.1139/b91-012.

Ghosh, B., Md, N.A. and Gantait, S. 2016. Response of rice under salinity stress: a review update. *Rice Research: Open Access 4*: 1–8.

Hosseini, S.J., Tahmasebi, S.Z. and Pirdashti, H. 2012. Screening of rice (*Oryza sativa* L.) genotypes for NaCl tolerance at early seedling stage. *International Journal of Agronomy and Plant Production 3*(8): 274–83.

Jan, Z., Ali, S., Sultan, T., Khan, M.J., Shah, Z. and Khan, F. 2018. Impact of different strains of cyanobacteria on rice crop growth and nutrients uptake under saline soil condition. *Sarhad Journal of Agriculture 34*(2): 450–58. http://dx.doi.org/10.17582/journal.sja/2018/34.2.450.458.

Johnson, N.C. 2010. Resource stoichiometry elucidates the structure and function of arbuscular mycorrhizas across scales. *New Phytologist 185*(3): 631–47. https://doi.org/10.1111/j.1469-8137.2009.03110.x.

Khaliq, A., Zia-ul-Haq, M., Ali, F., Aslam, F., Matloob, A., Navab, A. and Hussain, S. 2015. Salinity tolerance in wheat cultivars is related to enhanced activities of enzymatic antioxidants and reduced lipid peroxidation. *Clean-Soil, Air, Water 43*(8): 1248–58. https://doi.org/10.1002/clen.201400854.

Kumar, V., Shriram, V., Kavi Kishor, P.B., Jawali, N. and Shitole, M.G. 2010. Enhanced proline accumulation and salt stress tolerance of transgenic Indica rice by over-expressing P5CSF129A gene. *Plant Biotechnology Reports 4*(1): 37–48. https://doi.org/10.1007/s11816-009-0118-3.

Li, X. and Christie, P. 2001. Changes in soil solution Zn and pH and uptake of Zn by arbuscular mycorrhizal red clover in Zn-contaminated soil. *Chemosphere 42*(2): 201–7. https://doi.org/10.1016/S0045-6535(00)00126-0.

Mitra, D., Be, G.S., Khoshru, B., De Los Santos Villalobos, S., Belz, C., Chaudhary, P., Shahri, F.N., Djebaili, R., Adeyemi, N.O., El-Ballat, E.M. and El-Esawi, M.A., 2021c. Impacts of arbuscular mycorrhizal fungi on rice growth, development, and stress management with a particular emphasis on strigolactone effects on root development. *Communications in Soil Science and Plant Analysis 52*(14): 1591–621.

Mitra, D., Chaudhary, P., Verma, D., Khoshru, B., Senapati, A., Mahakur, B., Panneerselvam, P., Mohapatra, P.K.D. and Anđelković, S., 2021a. Bioinformatics' role in studying microbe-mediated biotic and abiotic stress tolerance. In: A. Kumar and S. Droby (Eds.), *Microbial Management of Plant Stresses* (pp. 203–219). Woodhead Publishing, Sawston.

Mitra, D., Djebaili, R., Pellegrini, M., Mahakur, B., Sarker, A., Chaudhary, P., Khoshru, B., Gallo, M.D., Kitouni, M., Barik, D.P. and Panneerselvam, P., 2021b. Arbuscular mycorrhizal symbiosis: plant growth improvement and induction of resistance under stressful conditions. *Journal of Plant Nutrition 44*(13): 1993–2028.

Muthuramalingam, P., Jeyasri, R., Rakkammal, K., Satish, L., Shamili, S., Karthikeyan, A., et al. 2022. Multi-omics and integrative approach towards understanding salinity tolerance in rice: a review. *Biology 11*(7): 1022. https://doi.org/10.3390/biology11071022.

Panneerselvam, P., Kumar, U., Senapati, A., Parameswaran, C., Anandan, A., Kumar, A., Jahan, A., Padhy, S.R. and Nayak, A.K., 2020. Influence of elevated CO_2 on arbuscular mycorrhizal fungal community elucidated using Illumina MiSeq platform in sub-humid tropical paddy soil. *Applied Soil Ecology 145*: 103344.

Panneerselvam, P., Kumar, U., Sugitha, T.C.K., Parameswaran, C., Sahoo, S., Binodh, A.K., Jahan, A. and Anandan, A., 2017. Arbuscular mycorrhizal fungi (AMF) for sustainable rice production. In: T.K. Adhya, B.B. Mishra, K. Annapurna, D.K. Verma and U. Kumar (Eds), *Advances in Soil Microbiology: Recent Trends and Future Prospects*, volume 2, Soil-Microbe-Plant Interaction (pp. 99–126). Springer, Singapore. https://doi.org/10.1007/978-981-10-7380-9_6.

Panneerselvam, P., Sahoo, S., Senapati, A., Kumar, U., Mitra, D., Parameswaran, C., Anandan, A., Kumar, A., Jahan, A. and Nayak, A.K., 2019. Understanding interaction effect of arbuscular mycorrhizal fungi in rice under elevated carbon dioxide conditions. *Journal of Basic Microbiology 59*(12): 1217–28.

Qin, W., Yan, H., Zou, B., Guo, R., Ci, D., Tang, Z., Zou, X., Zhang, X., Yu, X., Wang, Y. and Si, T., 2021. Arbuscular mycorrhizal fungi alleviate salinity stress in peanut: evidence from pot-grown and field experiments. *Food and Energy Security 10*(4): e314. https://doi.org/10.1002/fes3.314.

Shenoy, V.V. and Kalagudi, G.M. 2005. Enhancing plant phosphorus use efficiency for sustainable cropping. *Biotechnology Advances 23*(7–8): 501–13. https://doi.org/10.1016/j.biotechadv.2005.01.004.

Su, R., Zhang, Z., Chang, C., Peng, Q., Cheng, X., Pang, J., He, H. and Lambers, H. 2022. Interactive effects of phosphorus fertilization and salinity on plant growth, phosphorus and sodium status, and tartrate exudation by roots of two alfalfa cultivars. *Annals of Botany 129*(1): 53–64. https://doi.org/10.1093/aob/mcab124.

Tisarum, R., Theerawitaya, C., Samphumphuang, T., Polispitak, K., Thongpoem, P., Singh, H.P. and Cha-Um, S. 2020. Alleviation of salt stress in upland rice (*Oryza sativa* L. ssp. indica cv. Leum Pua) using arbuscular mycorrhizal fungi inoculation. *Frontiers in Plant Science 11*: 348. https://doi.org/10.3389/fpls.2020.00348.

Wang, C., Gu, Z., Cui, H., Zhu, H., Fu, S. and Yao, Q. 2015. Differences in arbuscular mycorrhizal fungal community composition in soils of three land use types in subtropical hilly area of Southern China. *PLoS One 10*(6): e0130983. https://doi.org/10.1371/journal.pone.0130983.

Wang, X., Yao, H., Wong, M.H. and Ye, Z. 2013. Dynamic changes in radial oxygen loss and iron plaque formation and their effects on Cd and As accumulation in rice (*Oryza sativa* L.). *Environmental Geochemistry and Health 35*(6): 779–88. https://doi.org/10.1007/s10653-013-9534-y.

11 Role of Arbuscular Mycorrhizal Fungi in the Alleviation of Heavy Metal Stress in Rice

E. Janeeshma
MES KEVEEYAM College

Joy M. Joel
University of Calicut

A.M. Shackira
Sir Syed College

Riya Johnson
University of Calicut

Thomas T. T. Dhanya
Carmel College

CONTENTS

11.1 INTRODUCTION

Arbuscular mycorrhizal fungi (AMF) are symbiotic fungi, which are associated with terrestrial plants including bryophytes, pteridophytes, gymnosperms, and angiosperms (Redecker et al., 2000; Read, 2002; Delaux, 2017; Chen et al.,

DOI: 10.1201/9781003354086-11

2018; Panneerselvam et al., 2017, 2019). These interdependent associations have several benefits to plant growth, including the enrichment of nutrition by increasing the surface area of roots and translocation of minerals, improved soil texture, increased organic substance decaying, etc. (Zou et al., 2016; Thirkell et al., 2017; Janeeshma and Puthur, 2020; Yutao et al., 2021). Moreover, the fungal association also alters the root morphology and leaf area, ultimately aiding the plant to significantly increase the biomass, which may help the flora to be equipped with adverse environmental circumstances like drought, salinity, flood, high/low temperature, and exposure to toxic metal ions (Gholamhoseini et al., 2013).

The reduction of the symptoms of heavy metal toxicity in plants with the aid of AMF is currently considered a potential biological tool as the AMF diminish the toxicity imparted by the toxic metal ions (Zhang et al., 2019; Dhalaria et al., 2020). The employment of AMF for the enhanced resistance to toxic metal ions in major agricultural crops including rice has paved the way to convene the global food security (Figure 11.1).

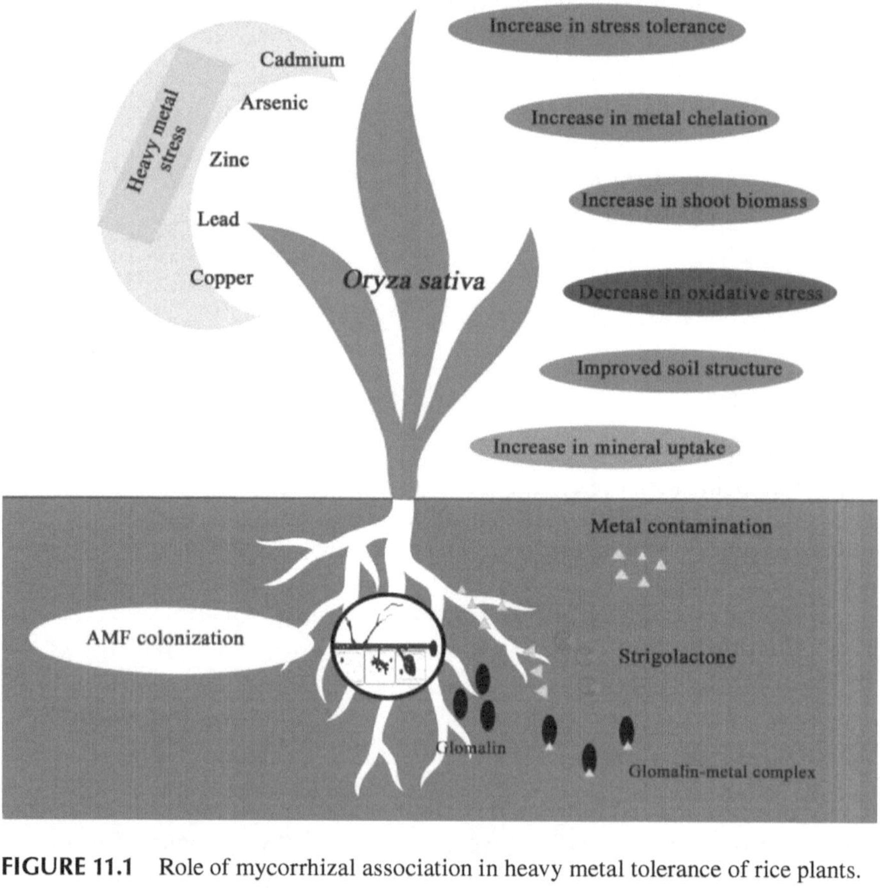

FIGURE 11.1 Role of mycorrhizal association in heavy metal tolerance of rice plants.

The presence of toxic metal ions in rice grains has turned out to be a severe concern worldwide as the consumption of these polluted grains may impose severe health hazards to human beings (Janeeshma et al., 2021a). So, it is critical to delimit the level of toxic metal ions from the surrounding soil surface to the grains, and ensuring the application of an eco-friendly approach is also crucial; otherwise, it may affect the structure and quality of the soil. Rice's roots are found to be highly effective in associating with AMF to mitigate metal toxicity either through degrading the pollutants by enzyme activity or through effectively accumulating/sequestering the metal ions (Mitra et al., 2022). Thus, the focus of this chapter is to document and emphasize the role of AMF in reducing heavy metal stress in rice plants, which may aid in improving sustainable agriculture practices as well as will be a promising tool in plant breeding approaches.

11.2 IMMOBILIZATION OF METAL IN THE MYCORRHIZOSPHERE

AMF, being one of the various rhizospheric microbial organisms to have intimately associated with plant roots, have augmented the plant's struggle to metal stress (Wu et al., 2016). Glomalin, the glycosylated protein of the fungal hyphal and spore wall, plays a significant role in the environment that includes the promotion of soil aggregation, carbon sequestration, and biostabilization of heavy and organic pollutants (Singh et al., 2016, 2022; Wang et al., 2012; Muchane et al., 2019).

The function of glomalin in remediating the toxic contaminant from soil by the AMF was first recognized by González-Chávez et al. (2004). Through a series of experiments, they extracted glomalins from polluted soil, hyphal structures, and *in vivo* sequestered copper in sorghum plants colonized by *Glomus mosseae*. It was found that 1 g of glomalin extracted from these samples had 1.6, 0.4, and 0.3 mg of Cu, indicating its immense potential for heavy metal sequestration without literally affecting the crop to be a pragmatic solution as far as rice crop is considered.

In an experiment by Zhang et al. (2005), two upland rice cultivars were grown in a blend of four different heavy metals, namely, copper, zinc, lead, and cadmium together with/without three different AMF, namely, *G. versiforme*, *G. mosseae*, and *G. diaphanum*, for studying the protective efficacy on the selected upland rice varieties from heavy metal stress. The three AMF deployed in the experiment could successfully colonize the two upland rice varieties but essentially in varying degrees of colonization. Significant differences in the uptake of heavy metals were observed in varying rice varieties. Among the three fungal species inoculated, *G. mosseae* imparted the most shielding effect from heavy metal stress, probably due to the excessive glomalin production, thereby greater heavy metal sequestration, in turn restricting the bioavailability of the heavy metals for the rice plants.

11.3 ACCUMULATION OF METAL IN FUNGAL STRUCTURES

It was observed that there was a significant surge in the production of arbuscules and vesicles with the increase in metal concentration and rise in the retention of metal ions in the roots of such plants. Similarly, in maize plants inoculated with *Gigaspora gigantea*, parallel results were observed such as an increase in the arbuscules and vesicles but a noteworthy rise in the translocation of nutrients such as nitrogen and phosphorous into the plants was also observed, which helps in the responsiveness of the plants to stress conditions. Various innovative approaches have been devised to study metal uptake, transfer, and transformation in AMF structures (Khalid et al., 2021).

AMF spores extracted from mine tailings concentrate calcium, chromium, iron, nickel, copper, bromine, thorium, yttrium, and uranium (Weiersbye et al., 1999). Cadmium (^{109}Cd) uptake by the roots and hyphae of *G. mosseae* in subterranean clover grown at high and low cadmium concentrations was studied in a pot experiment. It has been revealed that the mycorrhizal fungi successfully sequester the heavy metal inside the hyphal mass from the soil, restricting the translocation of cadmium into the shoots of the plant from roots (Joner and Leyval, 1997).

11.4 IMPORTANCE OF MYCORRHIZA IN STRESS TOLERANCE OF RICE UNDER HEAVY METAL TOXICITY

Plants with this connection to the mycorrhiza are healthier and can survive greater metal stress compared to plants without it. AMF penetrate plant roots, build a huge complex of extraradical mycelium in the rhizosphere, and assist their hosts in absorbing minerals and nutrients (Rivero et al., 2018). Arbuscular mycorrhizal connections are crucial for defending plants in metal-contaminated lands. Compared with plants in pristine environments, mycorrhizae-associated plants are more tolerant of heavy metal contamination.

Plants produce root exudates rich in flavonoids and strigolactones that draw mycorrhizal fungus. Additionally, fungi emit lipochitooligosacharide signals that plants can detect. This flow of signals is not hampered in some mycorrhizal associations by the persistent presence of toxic levels of heavy metals. This indicates that some mycorrhizal fungi can withstand extreme concentrations of such metals. The capability of AMF hyphae to attach to metals results in the immobilization of numerous heavy metals. Due to its binding capability, the transfer of harmful metal ions into the plant's parts is diminished (Brown and Wilkins, 1985).

According to Li et al. (2011), seeds infused with *G. geosporum* and *G. intraradices* may dramatically increase arsenic (As) tolerance, grain production, total plant weight, root phosphorus (P) absorption, and As uptake by the plant. However, there were no noticeable changes in grain As concentration. These findings revealed that, although increased P levels can significantly boost grain production, straw biomass, and root biomass, the content of As in grain is unaffected by these increases. Different species of AMF associated with the same host plant may have different patterns in phenological, nutritional, colonization, and gene expression responses (Feddermann et al., 2010). Plants may receive water and mineral nutrients from the AMF living in root cortical cells in exchange for carbon molecules (Hata et al., 2010).

Related to non-mycorrhizal-associated plants, plants treated with mycorrhiza are more tolerant to various stress conditions. As a result, plants with mycorrhizal associations exhibit toxic metal accumulation and excessive metal immobilization, without interfering with the mobilization of beneficial micro- and macronutrients, which can further help increase plant yield.

11.5 MITIGATION OF METAL-ELICITED OXIDATIVE STRESS BY MYCORRHIZATION

The presence of heavy metal stress produces the reactive oxygen species (ROS), which reduces the activities of the antioxidants (Janeeshma et al., 2021b,c). Because of reactions of radical displacement, various types of metallic ions are produced, which deplete the antioxidant activity (Bano and Ashfaq, 2013). The researchers reported that the AMF inoculation enhances the antioxidant synthesis, thereby increasing the antioxidant potential in heavy metal stress. The antioxidation system consists of antioxidants with non-enzymatic action such as ascorbate, glutathione, phenolics, proline, sugar, and enzymatic antioxidants like SOD (superoxide dismutase), CAT (catalase), peroxidase (glutathione peroxidase, GPX, and ascorbate peroxidase, APX), and GR (glutathione reductase) (Janeeshma and Puthur, 2020).

According to Tiwari and Lata (2018), the ROS production and antioxidant system were altered in mycorrhiza-associated rice under heavy metal stress. Under cadmium (Cd) stress conditions, the antioxidants including phenolics, GSH, and GPX increased in AMF-inoculated rice. The increase was shown in mycorrhizal-treated rice under high dosage of Cd (10 mg Cd/kg). The ROS level in rice plants is regulated by the antioxidant system in mycorrhiza. But the reactions of peroxidase, SOD, and CAT were reduced in AMF-inoculated rice under Cd stress. The genes encoding peroxidase, SOD, and CAT were downregulated during mycorrhizal association, which may reduce peroxidase, SOD, and CAT activities in mycorrhiza-associated rice plants under Cd stress. Additionally, AMF reduce the availability of Cd in the rhizosphere, which fixes in the cell wall that leads to reduced metal stress in plants that may reduce antioxidant activities (Li et al., 2020). Antioxidants with non-enzymatic action like proline, sugar, and phenolic content were increased in mycorrhiza-associated rice under Zn and Cd stress conditions. In mycorrhiza-associated rice leaves, proline, phenolic, and sugar contents were higher under Cd stress than under Zn stress (Janeeshma and Puthur, 2022). SOD and peroxidase activities were increased in AMF-inoculated rice plants under arsenite (As) stress conditions (Mitra et al., 2021, 2022).

11.6 CONCLUSION

The modification elicited by mycorrhization in the responses of rice plants under heavy metal stress was evaluated in this manuscript. The root hair formation, efficient antioxidant defense, glomalin exudation, metal sequestration in the fungal wall, etc. altogether aid to increasing the tolerance level of rice plants. These details should help further enhance the metal tolerance potential of rice plants.

Alt text file: Graphical representation is an image of rice plants showing arbuscular mycorrhizae associated with the roots of rice plants aid to tolerate heavy metal stress by reducing ROS production, increasing nutrient uptake, and improving soil structure.

REFERENCES

Bano, S. A., and Ashfaq, D. 2013. Role of mycorrhiza to reduce heavy metal stress. *Natural Science 5*:16–20.

Brown, M. T., and Wilkins, D. A. 1985. Zinc tolerance of mycorrhizal Betula. *New Phytologist 99*(1):101–106.

Chen, M., Arato, M., Borghi, L., Nouri, E., and Reinhardt, D. 2018. Beneficial services of arbuscular mycorrhizal fungi: From ecology to application. *Frontiers in Plant Science 9*:1270. doi: 10.3389/fpls.2018.01270.

Delaux, P. M. 2017. Comparative phylogenomics of symbiotic associations. *New Phytologist 213*:89–94. doi: 10.1111/nph.14161.

Dhalaria, R., Kumar, D., Kumar, H., et al. 2020. Arbuscular mycorrhizal fungi as potential agents in ameliorating heavy metal stress in plants. *Agronomy 6*:815. doi: 10.3390/agronomy10060815.

Feddermann, N., Finlay, R., Boller, T., and Elfstrand, M. 2010. Functional diversity in arbuscular mycorrhiza–the role of gene expression, phosphorous nutrition and symbiotic efficiency. *Fungal Ecology 3*:1–8.

Gholamhoseini, M., Ghalavand, A., Dolatabadian, A., Jamshidi, E., and Khodaei-Joghan, A., 2013. Effects of arbuscular mycorrhizal inoculation on growth, yield, nutrient uptake and irrigation water productivity of sunflowers grown under drought stress. *Agricultural Water Management, 117*:106–114.

González-Chávez, M. C., Carrillo-Gonzalez, R., Wright, S. F., and Nichols, K. A. 2004. The role of glomalin, a protein produced by arbuscular Mycorrhizal fungi, in sequestering potentially toxic elements. *Environment Pollution 130:*317–323. doi: 10.1016/j.envpol.2004.01.004.

Hata, S., Kobae, Y., and Banba, M. 2010. Interactions between plants and arbuscular mycorrhizal fungi. *International Review of Cell and Molecular Biology, 281*:1–48.

Janeeshma, E., Kalaji, H. M., and Puthur, J. T. 2021b. Differential responses in the photosynthetic efficiency of *Oryza sativa* and Zea mays on exposure to Cd and Zn toxicity. *Acta Physiologiae Plantarum 43:*1–16.

Janeeshma, E., and Puthur, J. T. 2020. Direct and indirect influence of arbuscular mycorrhizae on enhancing metal tolerance of plants. *Archives of Microbiology 202*:1–16.

Janeeshma, E., and Puthur, J. T. 2022. Physiological and metabolic dynamism in mycorrhizal and non-mycorrhizal *Oryza sativa* (var. Varsha) subjected to Zn and Cd toxicity: a comparative study. *Environmental Science and Pollution Research 11:*1–20.

Janeeshma, E., Puthur, J. T., and Ahmad, P. 2021a. Silicon distribution in leaves and roots of rice and maize in response to cadmium and zinc toxicity and the associated histological variations. *Physiologia Plantarum 173:*460–471.

Janeeshma, E., Rajan, V. K., and Puthur, J. T. 2021c. Spectral variations associated with anthocyanin accumulation; an apt tool to evaluate zinc stress in *Zea mays* L. *Chemistry and Ecology 37:*32–49.

Joner, E. J., and Leyval, C. 1997. Uptake of 109Cd by roots and hyphae of a *Glomus mosseae/Trifolium subterraneum* mycorrhiza from soil amended with high and low concentrations of cadmium. *New Phytologist 135*:353–360. doi: 10.1046/j.1469-8137.1997.00633.x.

Khalid, M., Ur-Rahman, S., Hassani, D., Hayat, K., Zhou, P., and Hui, N. 2021. Advances in fungal-assisted phytoremediation of heavy metals: a review. *Pedosphere 31*:475–495. doi: 10.1016/S1002-0160(20)60091-1.

Li, H., Chen, X. W., Wu, L., Luo, N., Huang, W. X., Mo, C. H., Li, Y. W., Xiang, L., Zhao, H. M., Cai, Q. Y., and Wong, M. H. 2020. Effects of arbuscular mycorrhizal fungi on redox homeostasis of rice under Cd stress. *Plant and Soil 455*:121–138.

Li, H., Ye, Z. H., Chan, W. F., Chen, X. W., Wu, F. Y., Wu, S. C., and Wong, M. H. 2011. Can arbuscular mycorrhizal fungi improve grain yield, As uptake and tolerance of rice grown under aerobic conditions? *Environmental Pollution 159*:2537–2545.

Mitra, D., Be, G. S., Khoshru, B., De Los Santos Villalobos, S., Belz, C., Chaudhary, P., Shahri, F. N., Djebaili, R., Adeyemi, N. O., El-Ballat, E. M., and El-Esawi, M. A. 2021. Impacts of arbuscular mycorrhizal fungi on rice growth, development, and stress management with a particular emphasis on strigolactone effects on root development. *Communications in Soil Science and Plant Analysis 52*:1591–1621.

Mitra, D., Saritha, B., Janeeshma, E., et al. 2022. Arbuscular mycorrhizal fungal association boosted the arsenic resistance in crops with special responsiveness to rice plant. *Environmental and Experimental Botany 193:*104681.

Muchane, M. N., Mirjam, M. P., Vanlauwe, B., Jefwa, J., and Kuyper, T. W. 2019. Impact of arbuscular mycorrhizal fungi and earthworms on soil aggregate stability, glomalin, and performance of pigeon pea, *Cajanus cajan*. *Soil Research 57*:53. doi: 10.1071/SR18096.

Panneerselvam, P., Kumar, U., Sugitha, T. C. K., Parameswaran, C., Sahoo, S., Binodh, A. K., Jahan, A., and Anandan, A. 2017. Arbuscular mycorrhizal fungi (AMF) for sustainable rice production. In *Advances in Soil Microbiology: Recent Trends and Future Prospects*, eds. T. K Adhya, B. Lal, B. Mohapatra, D. Paul, and S. Das, pp. 99–126. Springer, Singapore.

Panneerselvam, P., Sahoo, S., Senapati, A., Kumar, U., Mitra, D., Parameswaran, C., Anandan, A., Kumar, A., Jahan, A., and Nayak, A. K. 2019. Understanding interaction effect of arbuscular mycorrhizal fungi in rice under elevated carbon dioxide conditions. *Journal of Basic Microbiology 59*(12):1217–1228.

Read, D. J. 2002. An ecological point of view on arbuscular mycorrhiza research. In *Mycorrhizal Technology in Agriculture*, eds. S. Gianinazzi, H. Schüepp, J. M. Barea, K. Haselwandter, pp. 129–136. Birkhäuser, Basel. doi: 10.1007/978-3-0348-8117-3_11.

Redecker, D., Kodner, R., and Graham, L. E. 2000. Glomalean fungi from the Ordovician. *Science 289*:1920–1921. doi: 10.1126/science.289.5486.1920.

Rivero, J., Álvarez, D., Flors, V., Azcón-Aguilar, C., and Pozo, M. J. 2018. Root metabolic plasticity underlies functional diversity in mycorrhiza-enhanced stress tolerance in tomato. *New Phytologist 220:*1322–1336.

Singh, A. K., Rai, A., and Singh, N. 2016. Effect of long term land use systems on fractions of glomalin and soil organic carbon in the indo-gangetic plain. *Geoderma 277*:41–50. doi: 10.1016/j.geoderma.2016.05.004.

Singh, A. K., Xiai, Z., Chen, C., Wu, J., Yang, B., Zakari, S., Jiang, X. J., Singh, N., and Liu, W. 2022. The role of glomalin in mitigation of multiple soil degradation problems. *Critical Reviews in Environmental Science and Technology 52*:1604–1638. doi: 10.1080/10643389.2020.1862561.

Thirkell, T. J., Charters, M. D., Elliott, A. J., Sait, S. M., and Field, K. J. 2017. Are mycorrhizal fungi our sustainable saviours considerations for achieving food security. *Journal of Ecology 105*:921–929. doi: 10.1111/1365-2745.12788.

Tiwari, S., and Lata, C. 2018. Heavy metal stress, signaling, and tolerance due to plant-associated microbes: an overview. *Frontiers in Plant Science 9*:452.

Wang, F. Y., Wang, L., Yong, S. Z., Li, Y. J., and Song, M. Z. 2012. Effects of am inoculation and organic amendment, alone or in combination, on growth, p nutrition, and heavy-metal uptake of tobacco in Pb-Cd-contaminated soil. *Journal of Plant Growth Regulation 31*:549–559. doi: 10.1007/s00344-012-9265-9.

Weiersbye, I. M., Straker, C. J., and Przybylowicz, W. J. 1999. Micro-pixe mapping of elemental distribution in arbuscular mycorrhizal roots of the grass, *Cynodon dactylon*, from gold and uranium mine tailings. *Nuclear Instruments and Methods in Physics Research Section B: Beam Interactions with Materials and Atoms 158*:335–343. doi: 10.1016/S0168-583X(99)00467-X.

Wu, S., Zhang, X., Chen, B., Wu, Z., Li, T., Hu, Y., Sun, Y., and Wang, Y. 2016. Chromium immobilization by extraradical mycelium of arbuscular mycorrhiza contributes to plant chromium tolerance. *Environmental and Experimental Botany 122*:10–18. doi: 10.1016/j.envexpbot.2015.08.006.

Yutao, W, Xiaozhe, B., and Shaoshan, L. 2021. Effects of arbuscular mycorrhizal fungi on rice growth under different flooding and shading regimes. *Frontiers in Microbiology 12*:1–15.

Zhang, X. F., Hu, Z. H., and Yan, T. X. 2019. Arbuscular mycorrhizal fungi alleviate Cd phytotoxicity by altering Cd subcellular distribution and chemical forms in *Zea mays. Ecotoxicology and Environmental Safety 171*:352–360.

Zhang, X. H., Zhu, Y. G., Chen, B. D., Lin, A. J., Smith, S. E., and Smith, F. A. 2005. Arbuscular mycorrhizal fungi contribute to resistance of upland rice to combined metal contamination of soil. *Journal of Plant Nutrition 28*:2065–2077. doi: 10.1080/01904160500320871.

Zou, Y. N., Srivastava, A. K., and Wu, Q. S. 2016. Glomalin: a potential soil conditioner for perennial fruits. *International Journal of Agricultural Biology 18*:293–297. doi: 10.17957/IJAB/15.0085.

12 Arbuscular Mycorrhizal Fungi Association and Their Activation of Defense Response to Plant Protection

Khushneet Kaur, Kritika Gupta, and Shivangi Singh
Amity University

CONTENTS

12.1 INTRODUCTION

Plant-microbe interactions are common. Plant nutrition and resistance to adverse conditions are two common benefits. Arbuscular mycorrhizal fungi (AMF) of the subphylum Glomeromycotina are found in soil and are widely spread in agroforestry ecosystems (Fall et al., 2022). However, AMF associations and symbioses change plant physiology, resulting in improved mineral absorption and nutrition, as well as increased resistance/tolerance to biotic and abiotic stresses (Begum

DOI: 10.1201/9781003354086-12

et al., 2019; Panneersekvam et al., 2017). Furthermore, it has been stated that under natural circumstances, plants technically have mycorrhiza instead of roots; the roots of most flowering plants form symbiotic relationships with specific soil fungi (Azcón-Aguilar & Barea, 1997). AMF are the most prevalent type in typical cropping systems and in natural ecosystems, and they can be found in nearly all ecological situations. The numerous studies addressing mycorrhiza-induced resistance in various pathosystems show similar trends (Rodrigues & Rodrigues, 2019; van der Heijden, Martin, Selosse, & Sanders, 2015). AM fungal associations can be influenced by environmental conditions. Mycorrhizal plants have been frequently reported to have increased resistance to or tolerance to soilborne diseases.

According to various studies, ecosystems rely on AM symbioses to cycle nutrients, and the external mycorrhizal mycelium works with other soil organisms to form water-stable aggregates (Lehmann, Leifheit, & Rillig, 2017). Stronger plants because of mycorrhizal association may be better able to withstand stress due to an improved capacity for nitrogen uptake. However, AM symbioses can also give plants more focused boost in defense (improved tolerance and/or resistance to biotic and abiotic stresses).

Plants' responses to AMF entail the behavioral and physical activation of several defense systems (Ho-Plágaro & García-Garrido, 2022). The activation and management of these defenses have been postulated to play a role in the association's mutualistic state. Although there has been a revolution in understanding the activation and control of these defenses in relation to various host plants over the preceding decade, the same cannot be said for their role in supporting the rice production (Bahadur et al., 2019; Panneerselvam et al., 2019). In this context, the subsequent chapter analyses the potential of AMF for rice cultivation, with the molecular understanding and prospects to construct a self-sustaining rice production for the future.

12.1.1 Activation of Plant Defense by Mycorrhiza

According to experimental data, AM protects plants by combining several mechanisms that operate at various levels. The enhancement of plant nutrition and the symbiotic repair of pathogen damage are the most widely recognized explanations for mycorrhiza-induced protection. Mycorrhiza-induced resistance (MIR) is related to improved plant mineral nutrition, primarily phosphorus (Mitra et al., 2021). AMF improve plant nutrient content, which enhances plant photosynthetic rate and thus plant biomass. Specific plant responses and further activation of the plant defense system can be induced by mycorrhizal colonization in the root cortex. By recognizing effector proteins and microbe-associated molecular patterns (MAMP), plant receptor protein complexes can activate effector-triggered immunity (ETI) or MAMP-triggered immunity (MTI) (McDowell, 2019). Ruíz-Sánchez et al. (2011) demonstrated that the effector molecule secreted protein 7 (SP7) of the AM fungus *Glomus intraradices* encouraged the development of AM symbiosis by inhibiting plant immune responses. MTI in association with

AMF stimulates the synthesis of secondary metabolites and the generation of reactive oxygen.

Additionally, mycorrhizal colonization causes various quantitative and qualitative alterations in secondary metabolism in host plants. Multiple studies have shown that mycorrhizal infection causes the buildup of phenolic chemicals, alkaloids, terpenoids, and essential oils.

12.1.2 Phenolic Compounds

Phenolic compounds are the most extensive and ubiquitous category of allelochemicals, varying from simple phenolic acids to complex polymers such as tannins and lignin, and are responsible for plant chemical defense against microbial diseases and insect herbivores (Einhellig, 1986; Harborne, 1980). Plant inoculation with mycorrhizal fungus frequently leads to the buildup of phenolic chemicals.

Mycorrhizal inoculation, phenolic acids such as *trans*-caffeic, chlorogenic, *cis*-caffeic, protocatechuic, *cis*-p-coumaric, p-hydroxy benzoic, *trans*-p-coumaric, *trans*-ferulic, vanillic, and *cis*-ferulic acids are accumulated in groundnut (*Arachis hypogaea L.*) roots and shoots (Einhellig, 1986).

It is discovered that the isoflavonoid accumulation during VAM symbiosis. Cowpea (*Vigna unguiculata*) mycorrhizal stimulation with AMF *Glomus fasciculatum* inhibited pathogen colonization.

12.1.3 Phosphorous and Carbon

Typically, AMF enhance the host plant's absorption of phosphorus (P). Complex organic substances are broken down by soil enzymes into molecules bioavailable to plants and microorganisms. Their actions have a significant impact on nutrient cycling and the advancement of sustainable agriculture. It is demonstrated a high level of AMF colonization co-related with higher P uptake. AMF can transport greater fresh plant carbon to soil microorganisms than root exudation and can directly release up to 20%–30% of soil microbial carbon. AMF can boost bacterial activity by promoting gene expression. AMF also improve microbial development by stabilizing soil structure.

12.1.4 Abiotic Stress

The aim behind the use of AMF is to increase rice production and sustainability while minimizing environmental costs and boosting soil fertility. One objective is to increase the favorable effects of rice AMF interaction by identifying genetic traits that affect symbiotic result that might be used in breeding. Alleviation of abiotic stresses and harmful effects by AM symbiosis relies largely on better water and mineral nutrition in rice mycorrhized plants, which can further enhance physiological features such as root groundwater levels or photosynthetic capability.

12.1.5 Enzymatic Responses

AMF can significantly increase the activity of soil enzymes, releasing more bioavailable nutrients from complicated materials. Several studies indicate considerably varying outcomes regarding the influence of AMF on enzyme activity. AMF proliferation is correlated with enhanced enzymatic responses; consequently, a favorable environment for AMF growth can improve the influence that they have on the enzyme activity. Leifheit et al. (2014) conducted meta-analysis research to evaluate how AMF maintain soil structure and discovered that AMF prefer near-neutral soil pH that can further increase the functional performance of soil enzyme activity.

Previous research has shown that AMF can minimize nutrient absorption through subsoil (Rillig & Mummey, 2006) and avoid nutrient leakage. Meta-analysis found that AMF can increase soil enzyme production, thereby releasing more bioavailable nutrients. AMF reduced solely polyphenol oxidase activity in this investigation.

12.1.6 Molecular Responses

AMF symbiosis operates at the molecular level to influence the nutrient uptake pathways. The plant's perception of AMF infection triggers a series of host reactions, including the stimulation of host expression of genes.

AM fungus-colonized roots of diverse cereals had significantly greater transcriptional levels of 1-deoxy-D-xylulose 5-phosphate synthase (DXS) and 1-deoxy-D-xylulose 5-phosphate reductoisomerase (DXR), two important stages in the MEP pathway leading to apocarotenoids production (Walter, Fester, & Strack, 2000). Mycorrhizal formation, nodulation, and pathogen infection variably increase the gene expression of five chitinase classes in *Medicago truncatula* roots, with class III chitinases being selectively synthesized during arbuscular mycorrhizal infection (Salzer et al., 2000).

12.1.7 Signaling Responses

During AMF infection, a signaling cascade is initiated in the roots of host plants. This stimulation can be triggered by AM fungal compounds and occurs without any interaction with the symbionts. (Balestrini & Lanfranco, 2006). Jasmonic acid and its derivatives, known as jasmonates (JAs), are ubiquitous plant regulators and key plant defense signals against environmental stress. They are essential in mycorrhizal interactions. Furthermore, JA treatment enhanced root development in infected plants. Mostly in the field, rice could be attacked by a variety of pests, including herbivores, nematodes, and microbial infections such as the fungus *Magnaporthe oryzae* (the causative agent of the catastrophic rice blast), all of which trigger JA-signaling as an aspect of the plant defensive response. The effectively established that ET-induced systemic defense involves ET signaling and a substantial activation of JA biosynthesis and signaling genes, showing that the JA pathway is a critical defense route implicated in the rice root system RKN resistance. These findings suggest that JA is not necessary for AM colonization of rice, but rather high levels of JA in the roots limit AM growth, most probably through inducing defense.

12.2 ARBUSCULAR MYCORRHIZAL FUNGI (AMF) IN SUSTAINABLE RICE PRODUCTION

Given that rice is a staple meal consumed by most people worldwide and has a significant environmental impact, it is crucial to increase the sustainability of rice production (Okpiaifo et al., 2020). To maintain an agricultural system's production, capacity for regeneration, biological variety, vitality, and ability to function without endangering neighboring ecosystems, it must be managed and used sustainably. The use of AMF is a technique applied to produce rice sustainably. By providing rice with important nutrients in an accessible form without the use of unnecessary fertilizers, AMF play a critical role in nutrient management. Additionally, AMF increase the productivity in both flooded and unflooded rice (Panneerselvam et al., 2017).

12.2.1 Phosphorous and AMF

For the majority of their cellular and physiological processes, phosphorus is a nutrient that is necessary for the growth of living things. A phosphorous crisis on a worldwide scale is projected by 2050 as phosphorus progressively becomes a limiting nutrient. AMF plant symbiotic relationship provides a larger surface area for the nutrient availability as well as the uptake of phosphorous by translocation through hyphae or the mycelium (Panneerselvam et al., 2017). Iron and aluminum phosphates as well as rock phosphate are examples of poorly soluble P sources that mycorrhizal plants may absorb (Peterson & Massicotte, 2004). The AMF interaction is symbiotic, meaning that both parties gain from it. It helps the plant absorb water and nutrients from the soil's interphase, while the fungus uses the carbon the plant provides for growth, development, and other physiological processes. Although AMF are not colonization-specific, they occasionally show host preference. According to recent studies, land use intensification is reducing AMF diversity (Oehl et al., 2003). When rice plants are mycorrhized with *Glomus mosseae*, considerable amounts of insoluble P are mobilized from the soil and taken up by plants. AMF-inoculated rice plants enhance P absorption from weakly soluble phosphorous through direct or indirect mechanisms (Panneerselvam et al., 2017).

12.2.2 AMF and Disease Resistance

AMF are essential in the control of plant infections including *Phytophthora* species (Trotta et al., 1996) and *Rhizoctonia solani* (Yao, Tweddell, & Desilets, 2002). AMF have also been demonstrated to lessen bacterial infections in other crops (Chandanie, Kubota, & Hyakumachi, 2005). By creating secondary metabolites, parasitizing the plant pathogen, or inducing host resistance, AMF compete with plant pathogens for nutrition and space (Berg et al., 2007). AMF's biocontrol action has been explained by several processes, including biochemical changes, root system morphology, stress relief, and anatomical alterations (Hooker, Jaizme-Vega, & Atkinson, 1994).

12.2.3 Plant Growth-Regulating Substances by the AMF

AMF are also involved in the synthesis of vitamins like vitamin B and growth hormones including IAA, GA3, and cytokinin (Barea & Azcón-Aguilar, 1982). AMF consume hexose, which the mycelium then transforms into trehalose and glycogen. The carbon storage forms trehalose and glycogen may act as a buffer for the intracellular sugar content (Panneerselvam et al., 2017). The oxidative pentose phosphate pathway, which is also used to make nucleic acid, produces pentose. In intraradical mycelium, lipid production occurs (Pfeffer, Douds Jr, Bécard, & Shachar-Hill, 1999). The amount of cytokinin in the host plant increases because of VA mycorrhizal infection (Allen, Moore Jr, & Christensen, 1980).

12.2.4 AMF and Nutrient Management

By taking nitrate, phosphate, and ammonium from the soil, extraradical hyphae assimilate nutrients. The necessary material exchange between the fungus and the colonized plant occurs in arbuscules. The storing organ is the vesicle. The AMF have the biochemical potential to boost the availability of phosphorus and other dietary elements. This capacity may involve excreting chelating chemicals, root phosphatase activity, and acidification of the rhizosphere (Habte & Fox, 1993). These processes include root exudation pattern and pH alteration (Li & Christie, 2001). When inoculated with *Glomus etunicatum*, rice cultivar Pusa Basmati-1 cultivated in a Zn-deficient soil in a greenhouse experiment demonstrated that a higher intensity of colonization in rice may be attained by cultivating seedlings in P- and Zn-deficient soil in a nursery under aerobic circumstances (Purakayastha & Chhonkar, 2001).

AMF rice has a high quantity of glutathione when watered well. However, under drought stress conditions, glutathione levels in AMF rice have dropped. According to Ruíz-Sánchez et al. (2011), this drop in glutathione levels in AMF plants during dry conditions accelerates lipid peroxidation. Additionally, ascorbic and proline were enhanced during AMF vaccination, which made them even more potent defenses against the negative consequences of water scarcity. As a result, AMF are beneficial for rice plants as an acceptable and eco-friendly solution to enhance plant development and performance. According to research by Ruiz-Sánchez, Aroca, Muñoz, Polón, and Ruiz-Lozano (2010), the arbuscular mycorrhizal fungus *Glomus intraradices* improved the rice plant's growth response, photosynthetic efficiency, and antioxidative responses to drought stress.

REFERENCES

Allen, M. F., Moore Jr, T. S., & Christensen, M. (1980). Phytohormone changes in Bouteloua gracilis infected by vesicular–arbuscular mycorrhizae: I. Cytokinin increases in the host plant. *Canadian Journal of Botany, 58*(3), 371–374.

Azcón-Aguilar, C., & Barea, J. (1997). Arbuscular mycorrhizas and biological control of soil-borne plant pathogens–an overview of the mechanisms involved. *Mycorrhiza, 6*(6), 457–464.

Bahadur, A., Batool, A., Nasir, F., Jiang, S., Mingsen, Q., Zhang, Q., … Feng, H. (2019). Mechanistic insights into arbuscular mycorrhizal fungi-mediated drought stress tolerance in plants. *International Journal of Molecular Sciences, 20*(17), 4199. Retrieved from https://www.mdpi.com/1422-0067/20/17/4199.

Balestrini, R., & Lanfranco, L. (2006). Fungal and plant gene expression in arbuscular mycorrhizal symbiosis. *Mycorrhiza, 16*(8), 509–524.

Barea, J. M., & Azcón-Aguilar, C. (1982). Production of plant growth-regulating substances by the vesicular-arbuscular mycorrhizal fungus Glomus mosseae. *Applied and Environmental Microbiology, 43*(4), 810–813.

Begum, N., Qin, C., Ahanger, M. A., Raza, S., Khan, M. I., Ashraf, M., … Zhang, L. (2019). Role of arbuscular mycorrhizal fungi in plant growth regulation: implications in abiotic stress tolerance. *Frontiers in Plant Science, 10*, 1068.

Berg, G., Grosch, R., Scherwinski, K., Pujol, M., Badosa, E., Manceau, C., … Pinzani, P. (2007). Risk assessment for microbial antagonists: are there effects on. *Gesunde Pflanz, 59*, 107–117.

Chandanie, W., Kubota, M., & Hyakumachi, M. (2005). Interaction between arbuscular mycorrhizal fungus Glomus mosseae and plant growth promoting fungus Phoma sp. on their root colonization and growth promotion of cucumber (*Cucumis sativus* L.). *Mycoscience, 46*(3), 201–204.

Einhellig, F. A. (1986). Mechanisms and modes of action of allelochemicals. In: A. R. Putnam and C.-S. Tang (Eds.), The Science of Allelopathy (pp. 171–188). John Wiley & Sons, New York.

Fall, A. F., Nakabonge, G., Ssekandi, J., Founoune-Mboup, H., Apori, S. O., Ndiaye, A., … Ngom, K. (2022). Roles of arbuscular mycorrhizal fungi on soil fertility: contribution in the improvement of physical, chemical, and biological properties of the soil. *Fungi-Plant Interactions*. doi: 10.3389/ffunb.2022.723892.

Habte, M., & Fox, R. (1993). Effectiveness of VAM fungi in nonsterile soils before and after optimization of P in soil solution. *Plant and Soil, 151*(2), 219–226.

Harborne, J. (1980). Encyclopedia of plant phenolics. In: E.A. Bell & B.V. Charlwood (Eds.), Secondary Plant Products (pp. 329–402.). Springer, Berlin Heidelberg.

Ho-Plágaro, T., & García-Garrido, J. M. (2022). Molecular regulation of arbuscular mycorrhizal symbiosis. *International Journal of Molecular Sciences, 23*(11), 5960.

Hooker, J., Jaizme-Vega, M., & Atkinson, D. (1994). Biocontrol of plant pathogens using arbuscular mycorrhizal fungi. In: S. Gianinazzi & H. Schüepp (Eds.), *Impact of Arbuscular Mycorrhizas on Sustainable Agriculture and Natural Ecosystems* (pp. 191–200): Springer.

Lehmann, A., Leifheit, E., & Rillig, M. (2017). Mycorrhizas and soil aggregation. In: N. C. Johnson, C. Gehring, & J. Jansa (Eds.), *Mycorrhizal Mediation of Soil* (pp. 241–262): Elsevier, Amsterdam, Netherlands.

Li, X., & Christie, P. (2001). Changes in soil solution Zn and pH and uptake of Zn by arbuscular mycorrhizal red clover in Zn-contaminated soil. *Chemosphere, 42*(2), 201–207.

McDowell, J. M. (2019). Focus on activation, regulation, and evolution of MTI and ETI. *Molecular Plant-Microbe Interactions®, 32*(1), 5. doi: 10.1094/mpmi-11-18-0307-fi.

Mitra, D., Djebaili, R., Pellegrini, M., Mahakur, B., Sarker, A., Chaudhary, P., Khoshru, B., Gallo, M. D., Kitouni, M., Barik, D. P., & Panneerselvam, P. (2021). Arbuscular mycorrhizal symbiosis: plant growth improvement and induction of resistance under stressful conditions. *Journal of Plant Nutrition*, *44*(13), 1993–2028.

Oehl, F., Sieverding, E., Ineichen, K., Mäder, P., Boller, T., & Wiemken, A. (2003). Impact of land use intensity on the species diversity of arbuscular mycorrhizal fungi in agroecosystems of Central Europe. *Applied and Environmental Microbiology, 69*(5), 2816–2824.

Okpiaifo, G., Durand-Morat, A., West, G. H., Nalley, L. L., Nayga, R. M., & Wailes, E. J. (2020). Consumers' preferences for sustainable rice practices in Nigeria. *Global Food Security, 24*, 100345. doi: 10.1016/j.gfs.2019.100345.

Peterson, R. L., & Massicotte, H. B. (2004). Exploring structural definitions of mycorrhizas, with emphasis on nutrient-exchange interfaces. *Canadian Journal of Botany, 82*(8), 1074–1088.

Pfeffer, P. E., Douds Jr, D. D., Bécard, G., & Shachar-Hill, Y. (1999). Carbon uptake and the metabolism and transport of lipids in an arbuscular mycorrhiza. *Plant Physiology, 120*(2), 587–598.

Purakayastha, T., & Chhonkar, P. (2001). Influence of vesicular-arbuscular mycorrhizal fungi (*Glomus etunicatum* L.) on mobilization of zinc in wetland rice (*Oryza sativa* L.). *Biology and Fertility of Soils, 33*(4), 323–327.

Panneerselvam, P., Kumar, U., Sugitha, T. C. K., Parameswaran, C., Sahoo, S., Binodh, A. K., Jahan, A., & Anandan, A. (2017). Arbuscular mycorrhizal fungi (AMF) for sustainable rice production. In: T. K Adhya, B. Lal, B. Mohapatra, D. Paul, & S. Das (Eds.), *Advances in Soil Microbiology: Recent Trends and Future Prospects* (pp. 99–126). Springer, Singapore.

Panneerselvam, P., Sahoo, S., Senapati, A., Kumar, U., Mitra, D., Parameswaran, C., Anandan, A., Kumar, A., Jahan, A., & Nayak, A. K. (2019). Understanding interaction effect of arbuscular mycorrhizal fungi in rice under elevated carbon dioxide conditions. *Journal of Basic Microbiology, 59*(12), 1217–1228.

Rodrigues, K. M., & Rodrigues, B. F. (2019). Arbuscular Mycorrhizal (AM) Biotechnology and its applications.

Ruíz-Sánchez, M., Armada, E., Muñoz, Y., de Salamone, I. E. G., Aroca, R., Ruíz-Lozano, J. M., & Azcón, R. (2011). Azospirillum and arbuscular mycorrhizal colonization enhance rice growth and physiological traits under well-watered and drought conditions. *Journal of Plant Physiology, 168*(10), 1031–1037.

Ruiz-Sánchez, M., Aroca, R., Muñoz, Y., Polón, R., & Ruiz-Lozano, J. M. (2010). The arbuscular mycorrhizal symbiosis enhances the photosynthetic efficiency and the antioxidative response of rice plants subjected to drought stress. *Journal of Plant Physiology, 167*(11), 862–869.

Salzer, P., Bonanomi, A., Beyer, K., Vögeli-Lange, R., Aeschbacher, R. A., Lange, J., … Boller, T. (2000). Differential expression of eight chitinase genes in Medicago truncatula roots during mycorrhiza formation, nodulation, and pathogen infection. *Molecular Plant-Microbe Interactions, 13*(7), 763–777.

Trotta, A., Varese, G., Gnavi, E., Fusconi, A., Sampo, S., & Berta, G. (1996). Interactions between the soilborne root pathogenPhytophthora nicotianae var. parasitica and the arbuscular mycorrhizal fungusGlomus mosseae in tomato plants. *Plant and Soil, 185*(2), 199–209.

van der Heijden, M. G. A., Martin, F. M., Selosse, M.-A., & Sanders, I. R. (2015). Mycorrhizal ecology and evolution: the past, the present, and the future. *New Phytologist, 205*(4), 1406–1423. doi: 10.1111/nph.13288.

Walter, M. H., Fester, T., & Strack, D. (2000). Arbuscular mycorrhizal fungi induce the non-mevalonate methylerythritol phosphate pathway of isoprenoid biosynthesis correlated with accumulation of the 'yellow pigment'and other apocarotenoids. *The Plant Journal, 21*(6), 571–578.

Yao, M., Tweddell, R., & Desilets, H. (2002). Effect of two vesicular-arbuscular mycorrhizal fungi on the growth of micropropagated potato plantlets and on the extent of disease caused by Rhizoctonia solani. *Mycorrhiza, 12*(5), 235–242.

13 Management of Rice Phytopathogens through Arbuscular Mycorrhizal Fungi

Shraddha Bhaskar Sawant
Odisha University of Agriculture & Technology

Ankita Behura
Utkal University

S.R. Prabhukarthikeyan
ICAR-National Rice Research Institute

CONTENTS

13.1 INTRODUCTION

Rice (*Oryza sativa* L.) is a vital agricultural crop that feeds more than half of the world's populace (Campos-Soriano et al., 2012). Plant diseases play a significant role in the production of rice, having a substantial impact on earnings. In developing countries, it can result in annual production losses of 30%–50%, as well as food shortages and hunger (FAO, 2018). Given the crucial need for an eco-friendly and safe technique of controlling plant diseases, the use of biological control of plant pathogens is a compelling strategy for disease protection (Campo et al., 2020). In the biological approach to reducing the adverse impacts of phytopathogens, using various arbuscular mycorrhizal fungi (AMF) or AMF-based consortiums would be the best alternative (Poovarasan et al., 2013; Tisarum et al., 2020; Ding et al., 2019).

DOI: 10.1201/9781003354086-13

Rice is primarily grown in wetland settings, which have anaerobic circumstances that make AMF survival challenging. An increasing number of studies have discovered that AMF growth responses in rice plants range from positive to negative. The interaction between AMF and rice is impacted by soil pathogens. In paddy fields, *Rhizophagus irregularis* and *Funneliformis mosseae* are included in AMF species. Phosphate solubilization, nitrogen fixation, rice growth promotion, and pathogen suppression are among their synergistic stimulating activities. The AMF's biocontrol activity includes (i) parasitizing pathogens, (ii) increased competition for colonization sites and host photosynthates, (iii) increased competition for host photosynthates, (iv) secondary metabolite generation, (v) microbial community modification, (vi) boosting plant nutrient uptake and root system morphological modifications, and (vii) creating plant systemic resistance.

13.2 MOLECULAR INTERACTIONS OF AMF AND RICE PLANTS

When a plant interacts with any microbial partner, signal transduction occurs. The plant's roots release several volatile compounds, including strigolactone, which is sensitive to AMF and encourages spore germination as well as branch growth (Akiyama et al., 2005). A previously unrecognized part known as myc factor is formed during spore germination, which results in downstream reactions such as Ca^{2+} responses in the host, which causes membrane deterioration via peroxidation of lipid, inactivation of protein, and modification of DNA at the infection site, resulting in systemic resistance or hypersensitive response (HR). Plant hormones such as abscisic acid (ABA), gibberellins (GA), auxins, ethylene (ET), cytokinins (CK), and jasmonic acid (JA) and strigolactone play a role in the AM fungus-plant relationship, increasing systemic resistance and aiding in defense activation (Mitra et al., 2021; Gutjahr and Paszkowski, 2009). Throughout the AMF-host association, a complex mechanism results in an initial rise in SA, which progresses to systemic resistance. The accumulation of jasmonates in plant roots colonized with AM serves various roles, including enhanced flavonoid production, cytoskeleton remodeling, changes in root sink condition, and increased plant fitness (Hause et al., 2007). Chitin elicitors are released by ectomycorrhizal and AMF, which activates a defensive response (Salzer and Boller, 2000). Several studies have found that mycorrhizal colonization increases plant tolerance to harmful fungus (Pozo and Azcon-Aguilar, 2007; Garcia-Garrido and Ocampo, 2002; Song et al., 2015). Pathogenesis-related proteins (PR-Proteins) such as PR2 and PR3, and other genes related to defense, such as LOX, AOC, and PAL, are activated during pathogen attack in AMF-inoculated plants. During AM colonization, secondary metabolites increase. Primary and secondary metabolic processes in host plants are influenced by AM invasion (Schliemann et al., 2008). AM colonization changes the activities of both enzymes (e.g., catalase and superoxide dismutase). Phenylalanine ammonia lyase (PAL) is responsible for the production of phytoalexins and phenolic compounds in rice (Blilou et al., 2000).

13.3 AMF MECHANISMS IN BIOTIC STRESS MANAGEMENT

It was discovered that bacteria with similar physiological requirements compete for nutrition or place at the infection site in an ecological niche (Vos et al., 2014). Dehne (1982) studied the patterns of AMF colonization and root pathogens within similar host and observed that they frequently invade discrete cortical cells, signifying space rivalry. Mycorrhiza engage with soil in various ways, generating alterations in the microbial population in rhizosperic zones (Mitra et al., 2021). This association is beneficial in the control of plant pathogens. AMF help the host in two ways: directly by antagonism and indirectly by interacting with other plant growth or dangerous bacteria, which retards phytopathogens (Barea et al., 2002). PSB play an important function in plant disease control and can survive in soil for a longer period after being infected with AMF. Morphological studies have demonstrated that the AMF and plant roots undergo rapid changes during AM colonization. AM colonization causes nuclear mobility in plants which may arise because of changes in an organization of the plant cytoskeleton, as shown in plant-pathogen interactions (Kobayashi et al., 1992; Figure 13.1).

13.4 SIGNIFICANCE OF AMF IN THE MANAGEMENT OF RICE DISEASES

Bacterial leaf blight caused by *Xanthomonas oryza epvoryzae* (*Xoo*) is currently one of the most serious problems in lowland rice. This disease reduces yield by 30%–40% (Marlina and Fikrinda, 2020). Mycorrhiza influenced the incubation period, the length of the lesion, and severity of the disease. The maximum disease intensity was obtained in Impari 10 rice cultivar that were not provided mycorrhiza, 25.18%, and the lowest disease intensity was obtained in plants that were given mycorrhiza up to 15 and 20g per plant, 17.29%, and 16.51%, respectively (Marlina and Fikrinda, 2020). Rice brown spot disease (RBS), caused by *Cochliobolus miyabeanus*, is one of the most serious rice diseases, producing seedling blight when infected seeds are used. As a result, seedlings and older plants become weakened, and in the case of grain infection, grain quality and weight are reduced. The effectiveness of AMF alone or along with the bioagents *Pseudomonas fluorescens* (Pf) and *Trichoderma viride* in establishing systemic acquired resistance to rice brown spot disease. PR-proteins had also been identified (Shabana et al., 2008). The root colonization of the AM fungus *Glomus intraradices* is accompanied by the systemic activation of genes that regulate the rice defense response to the blast. OsNPR1, OsAP2, OsEREBP, and OsJAmyb are some genes and are upregulated in the mycorrhiza-inoculated rice plants (Campos-Soriano et al., 2012). *Rhizoctonia solani* (Basidiomycete), the pathogen responsible for rice sheath blight, is a soil-borne disease with a diverse host range. The symptoms appear on the stem near the water line as brown lesions following the tillering stage of rice (Dasgupta, 1992). According to Saleh et al. (2020), inoculating rice roots with AMF considerably reduced both indices of sheath blight damage, namely, lesion number and length, by increasing AMF spore density (Table 13.1).

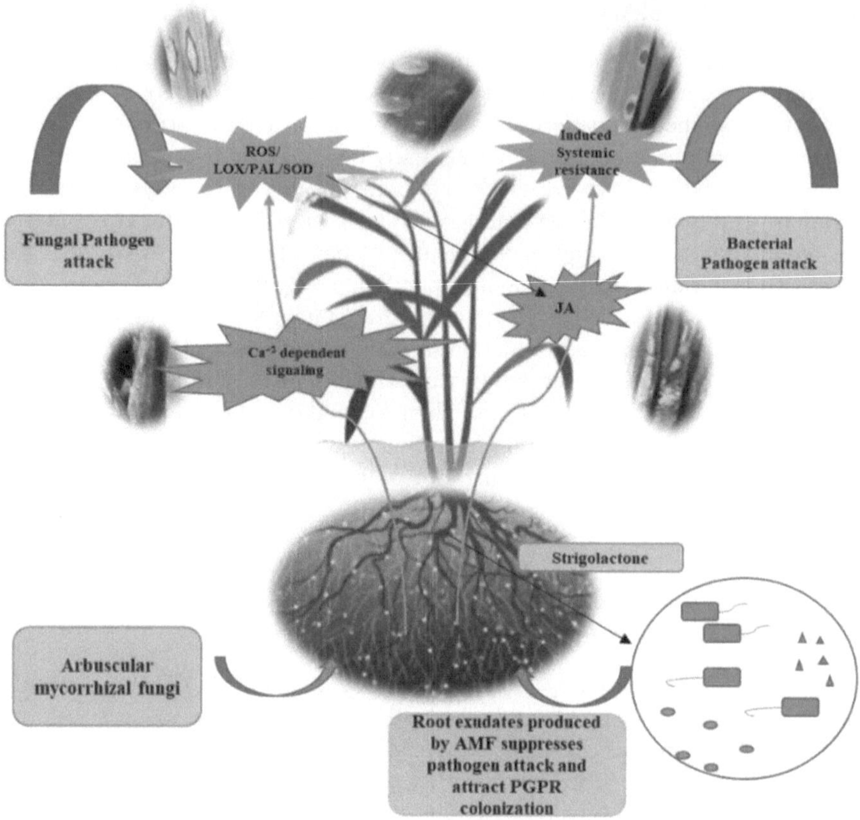

FIGURE 13.1 AMF colonization and activation of plant defense responses.

TABLE 13.1
Interactive Effect of AMF and Beneficial Microbes on Rice

Microbe Species	AMF Species	Effect on Rice	References
Bacillus megaterium	*Gigaspora* sp., *Acaulospora* sp.	Increased rice growth and yield	Prem Kumari et al. (2017)
Bacillus cepacia	*F. mosseae*	Increased P uptake, growth of rice	Hoseinzade et al. (2016)
Pseudomonas fluorescens	*R. irregularis*	Salinity decreased, grain yield	Radwan et al. (2008)
Psuedomonas synxantha	Single-spore AMF strain	Increased growth	Mäder et al. (2011)
Rhizoctonia solani	*G. aggregatum*	Decreased sheath blight infection	Baby and Manibhushanrao (1996)

13.5 CONCLUSION

Interactions between AMF and rice mycorrhizosphere microbes are among the most important and influential processes regulating the development and biotic stress management in rice. In the rice mycorrhizosphere, AMF invasion produces various metabolic alterations during pathogen infection, which activate plant defenses including cell wall modifications, increased secondary metabolism, and PR-protein accumulation. In the future, AMF must be used to manage rice crop diseases sustainably.

REFERENCES

Akiyama, K., Matsuzaki, K., and Hayashi, H. 2005. Plant sesquiterpenes induce hyphal branching in arbuscular mycorrhizal fungi. *Nature* 435:824–827.

Baby, U.I. and Manibhushanrao, K. 1996. Influence of organic amendments on arbuscular mycorrhizal fungi in relation to rice sheath blight disease. *Mycorrhiza* 6:201–206.

Barea, J.M., Azcon, R., and Azcon-Aguilar, C. 2002. Mycorrhizosphere interactions to improve plant fitness and soil quality. *Antonie Van Leeuwenhoek* 81:343–350.

Blilou, I., Ocampo, J.A., and García-Garrido, J.M. 2000. Induction of Ltp (lipid transfer protein) and pal (phenylalanine ammonia-lyase) gene expression in rice roots colonized by the arbuscular mycorrhizal fungus Glomus mosseae. *Journal of Experimental Botany* 51:1969–1977.

Campo, S., Martín-Cardoso, H., Olivé Pla, M., Catala-Forner, E., and Martínez Eixarch, M. 2020. Effect of root colonization by arbuscular mycorrhizal fungi on growth, productivity and blast resistance in rice. *Rice* 13:1–14.

Campos-Soriano, L., García-Martínez, J., and San Segundo, B. 2012. The arbuscular mycorrhizal symbiosis promotes the systemic induction of regulatory defence-related genes in rice leaves and confers resistance to pathogen infectionmpp_773. *Molecular Plant Pathology* 13(6): 579–592. doi: 10.1111/J.1364-3703.2011.00773.X.

Dasgupta, M.K. 1992. Rice sheath blight: The challenge continuous. In: U.S. Singh, A.N. Mukopadhayay, J. Kumar and H.S. Chanbe (eds.), *Plant Diseases of International Importance* (pp. 130–157, vol. I). Printice Hall, Hoboken, NJ.

Dehne, H.W. 1982. Interaction between vesicular-arbuscular mycorrhizal fungi and plant pathogens. *Phytopathology* 72:1115–1119.

Ding, L.J., Cui, H.L., Nie, S.A., Long, X.E., Duan, G.L., and Zhu, Y.G. 2019. Microbiomes inhabiting rice roots and rhizosphere. *FEMS Microbiology Ecology* 95:fiz040.

FAO. 2018. Rice market monitor. Available online: https://www.fao.org/3/I9243EN/i9243en.pdf (accessed on 25 April 2021).

Garcia-Garrido, J.M. and Ocampo, J.A. 2002. Regulation of the plant defence response in arbuscular mycorrhizal symbiosis. *Journal of Experimental Botany* 53:1377–1386.

Gutjahr, C. and Paszkowski, U. 2009. Weights in the balance: Jasmonic acid and salicylic acid signaling in root-biotroph interactions. *Molecular Plant-Microbe Interactions* 22:763–772.

Hause, B., Mrosk, C., Isayenkov, S., and Strack, D. 2007. Jasmonates in arbuscular mycorrhizal interactions. *Phytochemistry* 68:101–105.

Hoseinzade, H., Ardakani, M.R., Shahdi, A., Rahmani, H.A., Noormohammadi, G., and Miransari, M. 2016. Rice (*Oryza sativa* L.) nutrient management using mycorrhizal fungi and endophytic *Herbaspirillum seropedicae. Journal of Integrative Agriculture* 15:1385–1394.

Kobayashi, I., Kobayashi, Y., Yamaoka, N., and Kunoh, H. 1992. Recognition of a pathogen and a non-pathogen by barley coleoptile cells. III. Responses of microtubules and actin filaments in barley coleoptile cells to penetration attempts. *Canadian Journal of Botany* 70(9):1815–1823.

Mäder, P., Kaiser, F., Adholeya, A., Singh, R., Uppal, H.S., Sharma, A.K., Srivastava, R., Sahaid, V., Aragnoe, M., Wiemkenf, A., et al. 2011. Inoculation of root microorganisms for sustainable wheat–rice and wheat–black gram rotations in India. *Soil Biology and Biochemistry* 43:609–619.

Marlina, R. and Fikrinda, S. 2020. Effect of mycorrhizal fungi in controlling bacterial leaf disease in lowland rice caused by *Xanthomonas oryzae* pv oryzae bacteria. *Annals of the Romanian Society for Cell Biology*. Retrieved from https://annalsofrscb.ro/index.php/journal/article/view/36.

Mitra, D., Guerra Sierra, B.E., Khoshru, B., De Los Santos Villalobosd, S., Belze, C., Chaudhary, P., Shahrig, F.N., Djebaili, R., Adeyemi, N.O., El-Ballatk, E.M., El-Esawik, M.A., Moradic, S., Mondall, R., Senapati, A., Panneerselvam, P., and Das Mohapatra, P.K. 2021. Review impacts of arbuscular mycorrhizal fungi on rice growth, development, and stress management with a particular emphasis on strigolactone effects on root development. *Communications in Soil Science and Plant Analysis*. doi: 10.1080/00103624.2021.1892728.

Poovarasan, S., Mohandas, S., Paneerselvam, P., Saritha, B. and Ajay, K.M., 2013. Mycorrhizae colonizing actinomycetes promote plant growth and control bacterial blight disease of pomegranate (*Punica granatum* L. cv Bhagwa). *Crop Protection* 53:175–181.

Pozo, M.J. and Azcon-Aguilar, C. 2007. Unraveling mycorrhiza-induced resistance. *Current Opinion in Plant Biology* 10:393–398.

Prem Kumari, S.M., and Prabina, B.J. 2017. Impact of the mixed consortium of indigenous arbuscular mycorrhizal fungi (AMF) on the growth and yield of rice (*Oryza sativa* L.) under the system of rice intensification (SRI). *International Journal of Environment, Agriculture and Biotechnology* 2:238743.

Radwan, F.I., El-Seoud, A.I., and El-Ham, A.B. 2008. Response of two rice cultivars to Blue Green Algae, a-mycorrhizae inoculation and mineral nitrogen fertilizer. *Middle Eastern and Russian Journal of Plant Science and Biotechnology* 2:29–34.

Salzer, P. and Boller, T. 2000. Elicitor-induced reactions in mycorrhizas and their suppression. In: G.K. Podila and D.D. Douds (eds.), *Current Advances in Mycorrhizas Research* (pp. 1–10). APS Press, St. Paul, MN.

Schliemann, W., Ammer, C., and Strack, D. 2008. Metabolite profiling of mycorrhizal roots of Medicago truncatula. *Phytochemistry* 69:112–146.

Shabana, Y.M., Abdel-Fattah, G.M., Ismail, A.E., and Rashad, Y.M. 2008. Control of brown spot pathogen of rice (*Bipolaris oryzae*) using some phenolic antioxidants. *Brazilian Journal of Microbiology* 39(3):438–444. doi: 10.1590/S1517-83822008000300006.

Song, Y., Chen, D., Lu, K., Sun, Z., and Zeng, R. 2015. Enhanced tomato disease resistance primed by arbuscular mycorrhizal fungus. *Frontiers in Plant Science* 6:786.

Tisarum, R., Theerawitaya, C., Samphumphuang, T., Polispitak, K., Thongpoem, P., Singh, H.P., and Cha-Um, S. 2020. Alleviation of salt stress in upland rice (*Oryza sativa* L. ssp. indica cv. Leum Pua) using arbuscular mycorrhizal fungi inoculation. *Frontiers in Plant Science* 11:348.

Vos, C.M., Yang, Y., DeConinck, B., and Cammue, B.P.A. 2014. Fungal (–like) biocontrol organisms in tomato disease control. *Biological Control* 74:65–81.

14 Role of Arbuscular Mycorrhizal Fungi in Rice Insect and Nematode Management

G. Guru-Pirasanna-Pandi, Swagatika Sahoo, and Sampriti Mohanty
ICAR-National Rice Research Institute

CONTENTS

14.1 INTRODUCTION

For approximately half of the world's population, rice is a staple food, which is crucial for ensuring global food security (Muthayya et al., 2014). Its production is being constrained by numerous herbivores (insects, pathogen, nematode, etc.) (Matteson, 2000). To meet the growing demand of crop production, there is rampant use of chemicals in agriculture. The use of pesticides in agriculture substantially lowers crop losses by protecting the plant against various pests *viz.* fungi, bacteria, plant-parasite nematodes, and insects. But, overuse of these chemicals has adverse consequences to human health and the environment (Wani et al., 2017). Thus, environmental-friendly methods of plant protection are required in order to lessen the adverse effects of pesticide use. Eighty percent of terrestrial plant species have mutualistic associations between their roots called arbuscular mycorrhizal fungi (AMF). AMF are the soil organisms that are known to boost crop production by supporting plants to withstand a variety of abiotic and biotic stresses (Vannette and Hunter, 2009; Roger et al., 2013). The effects of symbiosis against different pests of rice and other crops have been described.

DOI: 10.1201/9781003354086-14

14.2 MECHANISM OF INTERACTION BETWEEN AMF AND INSECT

Resistance and tolerance are two primary crop protection techniques of plants for minimizing the negative effects of herbivore/pest attack (Mitchell et al., 2016). AMF can alter pairwise interactions, which is known to result in a variety of physiological, morphological, and biochemical alterations in host plants affecting plant resistance and tolerance as reported by Yang et al. (2014). It has been evidenced from several studies that association with AMF can affect plant defense responses or plant quality in a way that increases or decreases herbivore resistance (Cosme et al., 2011; Koricheva et al., 2009). Mycorrhizal colonization has a positive effect on mono, oligophagous chewing herbivores, and sucking insects, whereas the performance of polyphagous chewers and gall-forming insects was reduced and no effect was observed on leaf-mining insects (Koricheva et al., 2009). By changing the plant's nutrient status, photosynthetic activity, and/or growth, plant associations with soil microorganisms (AMF) may also provide tolerance against herbivores (Gange, 2001; Johnson et al., 2016). Factors like species of AMF (Koricheva et al., 2009), the soil type on which plants are grown (Bernaola and Stout, 2019), insect type—generalist/specialist (Hartley and Gange, 2009), number of mychorhizal species (Currie et al., 2011), and relatedness among AMF (Roger et al., 2013) influence their interaction with the host plant and with the herbivore.

Rice plants are affected by a diverse complex of pests viz. including stem borer, leaf folder, plant hoppers, midge, bugs, root weevil, and several other caterpillar pests (Jena et al., 2018). Cereal crops show high responsiveness to AMF (Sawers et al., 2008). Studies have been done on the use of AMF as insect-pest control for various crop systems but there are few information about their use against rice insect-pest in particular (Table 14.1). Associations with AMF in rice lead to modifications in competitiveness, ecotype specificity, functional diversity, uptake of nutrients, growth, and gene expression in plants. Although research on rice's interactions with AMF and herbivores has largely been conducted in isolation, few studies are there on interactions between herbivores, AMF, and rice crops (Bernaola and Stout, 2021). Environmental factors may have impact on these tripartite interactions.

14.3 MECHANISM OF INTERACTION BETWEEN AMF AND PLANT-PARASITIC NEMATODES

Numerous species of plant-parasitic nematodes (PPNs) have an impact on a number of economically significant agricultural crops. Among the most destructive nematodes for rice are the sedentary root-knot nematode *Meloidogyne graminicola* and the migratory root-lesion nematode *Pratylenchus zeae* (Kyndt et al., 2014). *Hirschmanniella oryzae* is another important nematode affecting rice crop (Seenivasan et al., 2010). PPN and AMF frequently coexist in the rhizosphere, colonizing the same area in host plant roots (Wani et al., 2017). Elsen et al. (2008)

TABLE 14.1
Interaction between AMF and Insects in Different Crops with Their Mechanism of Action

Target Organism	Host Crop	AMF Species	Effect on Host Crop	Effect on Insect	Mechanism	References
Rice water weevil (*Lissorhoptrus oryzophilus*)	Rice (*Oryzae sativa*)	*Rhizophagus intraradices*	(+) ve	(+) ve	More of nitrogen and phosphorus in the leaves and roots of rice plants with AMF inoculation, root weevils lay more eggs on those plants	Cosme et al. (2011)
Rice water weevil (*L. oryzophilus*) and Fall armyworm (*Spodoptera frugiperda*)	Rice (*O. sativa*)	Mixture of AMF species (*Funneliformis mosseae*, *Rhizophagus irregularis*, *Glomus deserticola*, *G. microaggregatum*, *R. fasciculatum*, and *Sclerocystis dussii*)	(+) ve	(+) ve	Increased root and shoot biomass in mycorrhizae colonized plants and increased densities in larvae of both the insects	Bernaola et al. (2018) and Bernaola and Stout (2019)
Rice water weevil (*L. oryzophilus*)	Rice (*O. sativa*)	Mixture of four AMF species: *R. irregularis*, *Glomus aggregatum*, *F. mosseae*, and *Claroideoglomus etunicatum*	(+) ve	(–) ve	AMF treatment on rice seed promotes growth of plants, yield, and root injury tolerance	Bernaola and Stout (2021)
Fall armyworm (*S. frugiperda*)	Rice (*O. sativa*)	*Glomus intraradices*	(–) ve	(+) ve	The plant biomass reduced and slight decline in plant defense enzymes (peroxidase and Anti-oxidant-polyphenol oxidase) in AMF inoculated injured plants	Bhavanam and Stout (2021)
Beet armyworm (*Spodoptera exigua*)	Tomato (*Solanum lycopersicum*)	*Funneliformis mosseae*	(+) ve	(–) ve	Upon insect feeding AMF inoculated plants produce more alkaloids, fatty acid derivatives as a local and systemic response	Rivero et al. (2021)

mentioned that in many plant systems, AMF increased host tolerance or resistance and developed systemic resistance to combat plant-parasitic nematodes. There is competition by AMF and PPN for the same site in the rhizosphere, where AMF colonize faster (Wani et al., 2017). Plants having enhanced nutrient concentration tolerate denser PPN populations in their roots because AMF boost the absorption of water and mineral in the host plant. Schouteden et al. (2015) noted that after the application of mycorrhizae, the root exudates are altered such that they reduce the penetration of nematodes into mycorrhizal plants and temporarily paralyze nematodes.

AMF can have both positive and negative effects on the nematode pests, AMF increase the root branching of host crop which can help the host plant to tolerate the damage by PPN, and at the same time, there are chances that the potential infection sites of nematodes might increase (Schouteden et al., 2015). Using biological agents, such as AMF, to manage PPN is a more environmentally friendly option (Bajaj et al., 2017). The defensive effect AMF against PPN was studied in various crops but few are reported on rice (Table 14.2).

14.4 EFFECTS OF AMF INOCULATION ON PLANT DEFENSE

The signaling pathways involved in defense are triggered when plants interact with herbivores, pathogens, nematodes, and beneficial organisms (Figure 14.1). They stimulate the plant hormones *viz.* ethylene (ET), jasmonic acid (JA), and salicylic acid (SA) (Feys and Parker, 2000). PAMP-triggered immunity (PTI) is triggered when a microbe or pest is detected at the cell surface and typically prevents infection before the pest establishes on the plant. The SA-responsive genes are regulated downstream of NPR1 and lead to upregulation of pathogenesis-related (PR) proteins. Systemic acquired resistance (SAR) responsive genes are often used to indicate a salicylate mediated plant defense response.

SAR is distinct from induced systemic resistance (ISR), which is produced when a plant interacts with helpful microbes. Additionally, this marks the start of the signaling pathways for ethylene and/or jasmonate, which are activated following the initial host-beneficial organism recognition (Feys and Parker, 2000; Derksen et al., 2013). So, here comes the application of AMF (beneficial organism), which can help in regulating such defense mechanism in plants against different pest attacks. JA concentrations increase at the site of infection and activate the plant defensive genes, also proteinase inhibitors, vegetative storage proteins, and enzymes synthesizing phytoalexin, thionins, and defensins (Hunter, 2000; Devoto and Turner, 2005). Mycorrhizal colonization "primes" plant systemic defenses combating pest attack by inducing JA levels (Conrath et al., 2006). AMF cause JA-induced responses, which in turn activates ISR, which is an effective defense against herbivory (Vannette and Hunter, 2009). AMF may affect their growth, developmental rate, and fecundity rate of insects/nematodes (Koricheva et al., 2009). According to the vast majority of studies, generalist chewing insects perform worse when AMF invade crops (Koricheva et al., 2009). According to

TABLE 14.2
Effect of Interaction between AMF and Plant-Parasitic Nematodes in Different Crops

Nematodes	Host Crop	AMF Species	Effect on Host Crop	Effect on Nematode	Mechanism	References
Root-knot nematode (*Meloidogyne graminicola*)	Rice (*O. sativa*)	Not specified	(+) ve	(−) ve	Fungi in the rhizosphere and on the surface of plants aided in promoting plant growth and improved nematode resistance	Le et al. (2009)
Lesion nematode (*Pratylenchus neglectus*)	Wheat (*Triticum aestivum*)	Four species: *Claroideoglomus entunicatum*, *Funneliformis coronatum*, *Rhizophagus irregularis*, and *F. mosseae*	(−) ve	(+) ve	Plant biomass was decreased by AM fungi, which also suppressed root defense compounds (benzoxazinoid glucoside) linked to plant resistance. However, plant nutrition concentration was increased	Frew et al. (2018)
Burrowing nematodes (*Radopholus similis*) and lesion nematode (*P. coffeae*)	Banana (*Musa* sp.)	*Glomus intraradices*	(+) ve	(−) ve	Induced systemic resistance (ISR) reduced the root necrosis	Elsen et al. (2008)
False root-knot nematode (*Nacobbus aberrans*)	Tomato (*S. lycopersicum*)	*Glomus intraradices*	(+) ve	(−) ve	Number of galls got reduced due to AMF induction	Lax et al. (2011)

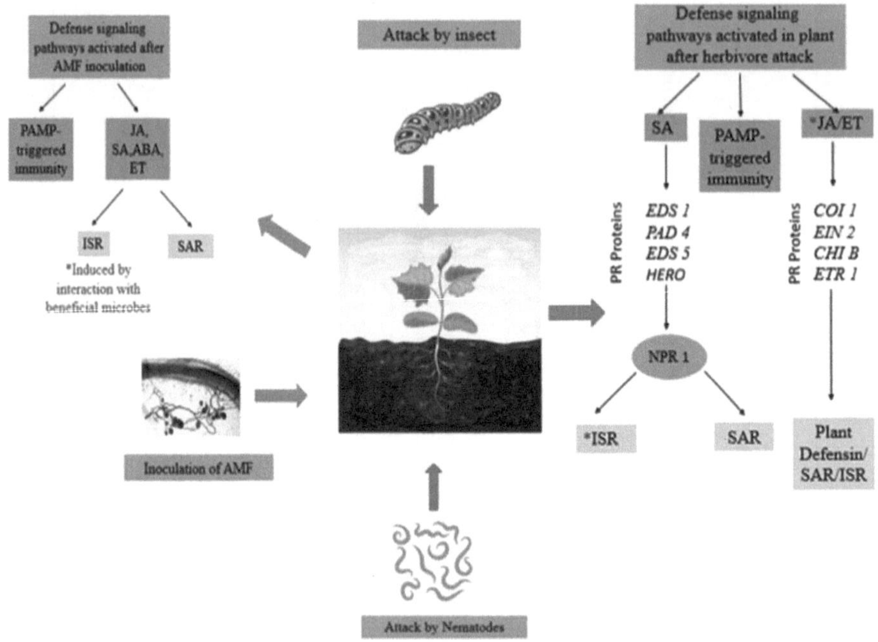

FIGURE 14.1 Induction of defense mechanism (signaling pathway) in plant triggered by various pest attack and inoculation of AMF (JA, jasmonic acid; SA, salicylic acid; ISR, induced systemic resistance; SAR, systemic acquired resistance; PR, protein-pathogenesis-related protein; ABA, abscisic acid; PAMP, pathogen-associated molecular patterns).

Hunter (2000), attack by phloem-feeding sucking pests occasionally fails to cause JA-induced responses, rather causes salicylic acid responses.

AMF also elicit some indirect defenses to crops. The quantity and type of volatile organic compounds (VOCs) may be affected by AMF, which would help attract natural enemies (Vannette and Hunter, 2009). These studies collectively imply that AMF can mediate alterations in plant to show resistance against different pests.

14.5 CONCLUSION

Arbuscular mycorrhizal fungi have been found to provide protection against different crop pests including, insects, viruses, phytoplasmas, bacteria, fungi, and nematodes that damage crop plants. Increasing rice production and fostering efficient and sustainable management of rice pests require an understanding of the interactions between AMF and the rice plant and also the alteration in plant system against various biotic stresses. Since the mechanisms underlying the responses by plants are still unknown, research on the effectiveness of

AM-enhanced plant resistance and its commercial exploitation as a biofertilizer and bioprotectant is ongoing.

REFERENCES

Bajaj, R., Prasad, R., Varma, A., and Bushley, K.E. 2017. The role of arbuscular mycorrhizal fungi and the mycorrhizal-like fungus *Piriformospora indica* in biocontrol of plant parasitic nematodes. In: Varma, A., Prasad, R., and Tuteja, N. (eds.), *Mycorrhiza- Eco-Physiology, Secondary Metabolites, Nanomaterials.* Springer, Cham, pp. 43–56.

Bernaola, L., Cosme, M., Schneider, R.W., and Stout, M. 2018. Belowground inoculation with arbuscular mycorrhizal fungi increases local and systemic susceptibility of rice plants to different pest organisms. *Frontiers in Plant Science,* 9:747.

Bernaola, L., and Stout, M.J. 2019. Effects of arbuscular mycorrhizal fungi on rice-herbivore interactions are soil-dependent. *Scintific Reports*, 9:14037.

Bernaola, L., and Stout, M.J. 2021. The effect of mycorrhizal seed treatments on rice growth, yield, and tolerance to insect herbivores. *Journal of Pest Science*, 94(2):375–392.

Bhavanam, S., and Stout, M.J. 2021. Assessment of silicon-and mycorrhizae-mediated constitutive and induced systemic resistance in rice, *Oryza sativa* L., against the fall armyworm, *Spodoptera frugiperda* Smith. *Plants*, 10(10):2126.

Conrath, U., Beckers, G.J., Flors, V., García-Agustín, P., Jakab, G., Mauch, F., Newman, M.A., Pieterse, C.M., Poinssot, B., Pozo, M.J., and Pugin, A. 2006. Priming: getting ready for battle. *Molecular Plant-Microbe Interactions*, 19(10):1062–1071.

Cosme, M., Stout, M.J, and Wurst, S. 2011. Effect of arbuscular mycorrhizal fungi (*Glomus intraradices*) on the oviposition of rice water weevil (*Lissorhoptrus oryzophilus*). *Mycorrhiza*, 21:651–658.

Currie, A.F., Murray, P.J., and Gange, A.C. 2011. Is a specialist root-feeding insect affected by arbuscular mycorrhizal fungi? *Applied Soil Ecology*, 47:77–83.

Derksen, H., Rampitsch, C., and Daayf, F. 2013. Signaling cross-talk in plant disease resistance. *Plant Science,* 207:79–87.

Devoto, A., and Turner, J.G. 2005. Jasmonate-regulated Arabidopsis stress signalling network. *Physiologia Plantarum,* 123:161–172.

Elsen, A., Gervacio, D., Swennen, R., and Waele, D.D. 2008. AMF-induced biocontrol against plant parasitic nematodes in Musa sp. a systemic effect. *Mycorrhiza*, 18:251–256.

Feys, B.J., and Parker, J.E., 2000. Interplay of signaling pathways in plant disease resistance. *Trends in Genetics*, 16(10):449–455.

Frew, A., Powell, J.R., Glauser, G., Bennett, A.E., and Johnson, S.N. 2018. Mycorrhizal fungi enhance nutrient uptake but disarm defences in plant roots, promoting plant-parasitic nematode populations. *Soil Biology and Biochemistry*, 126:123–132.

Gange, A.C. 2001. Species-specific responses of a root- and shoot feeding insect to arbuscular mycorrhizal colonization of its host plant. *New Phytologist*, 150:611–618.

Hartley, S.E. and Gange, A.C. 2009. Impacts of plant symbiotic fungi on insect herbivores: mutualism in a multitrophic context. *Annual Review of Entomology*, 54:323–342.

Hunter, M.D. 2000. Mixed signals and cross-talk: interactions between plants, insect herbivores, and plant pathogens. *Agricultural and Forest Entomology,* **2**, 155–160.

Jena, M., Adak, T., Rath, P.C., Gowda, G.B., Patil, N.B., Prasanthi, G., and Mohapatra, S.D. 2018. Paradigm shift of insect pests in rice ecosystem and their management strategy. *ORYZA-An International Journal on Rice*, 55:82–89.

Johnson, S.N., Erb, M., and Hartley, S.E. 2016. Roots under attack: contrasting plant responses to below- and aboveground insect herbivory. *New Phytologist*, 210:413–418.

Koricheva, J., Gange, A.C., and Jones, T. 2009. Effects of mycorrhizal fungi on insect herbivores: a meta-analysis. *Ecology,* 90:2088–2097.

Kyndt, T., Fernandez, D., and Gheysen, G. 2014. Plant-parasitic nematode infections in rice: molecular and cellular insights. *Annual Review of Phytopathology*, 52:135–153.

Lax, P., Becerra, A.G., Soteras, F., Cabello, M., and Douchet, M.E. 2011. Effect of the arbuscular mycorrhizal fungus *Glomus intraradices* on the false root-knot nematode *Nacobbus aberrans* in tomato plants. *Biology and Fertility of Soils*, 47:591–597.

Le, H.T.T., Padgham, J.L., and Sikora, R.A. 2009. Biological control of the rice root-knot nematode *Meloidogyne graminicola* on rice, using endophytic and rhizosphere fungi. *International Journal of Pest Management,* 55(1):31–36.

Matteson, P. C. 2000. Insect pest management in tropical Asian irrigated rice. *Annual Review of Entomology*, 45:549–574.

Mitchell, C., Brennan, R.M., Graham, J., and Karley, A.J. 2016. Plant defense against herbivorous pests: exploiting resistance and tolerance traits for sustainable crop protection. *Frontiers in Plant Science*, 7:1132.

Muthayya, S., Sugimoto, J.D., Montgomery, S., and Maberly, G.F. 2014. An overview of global rice production, supply, trade, and consumption. *Annals of New York Academy of Sciences*, 1324:7–14.

Rivero, J., Lidoy, J., Llopis-Giménez, Á., Herrero, S., Flors, V., and Pozo, M.J. 2021. Mycorrhizal symbiosis primes the accumulation of antiherbivore compounds and enhances herbivore mortality in tomato. *Journal of Experimental Botany*, 72(13):5038–5050.

Roger, A., Colard, A., Angelard, C. and Sanders, I.R. 2013. Relatedness among arbuscular mycorrhizal fungi drives plant growth and intraspecific fungal coexistence. *The ISME journal*, 7(11): 2137–2146.

Sawers, R.J.H., Gutjahr, C., and Paszkowski, U. 2008. Cereal mycorrhiza: an ancient symbiosis in modern agriculture. *Trends in Plant Science*, 13:1360–1385.

Schouteden, N., De Waele, D., Panis, B., and Vos, C.M. 2015. Arbuscular mycorrhizal fungi for the biocontrol of plant-parasitic nematodes: a review of the mechanisms involved. *Frontiers in Microbiology,* 6:1280.

Seenivasan, N., David, P.M.M., and Vivekanandan, P. 2010. Population dynamics of rice nematodes under a system of rice intensification. *Nematologia Mediterranea,* 38:159–163.

Vannette, R.L., and Hunter, M.D. 2009. Mycorrhizal fungi as mediators of defence against insect pests in agricultural systems. *Agricultural and Forest Entomology*, 11:351–358.

Wani, K.A., Manzoor, J., Shuab, R., and Lone, R. 2017. Arbuscular mycorrhizal fungi as biocontrol agents for parasitic nematodes in plants. In: Varma, A., Prasad, R., and Tuteja, N. (eds.), *Mycorrhiza-Nutrient Uptake, Biocontrol, Ecorestoration.* Springer, Cham, pp. 195–210.

Yang, H., Dai, Y., Wang, X., Zhang, Q., Zhu, L., and Bian, X. 2014. Meta-analysis of interactions between arbuscular mycorrhizal fungi and biotic stressors of plants. *The Scientific World Journal*, 24:1–7.

15 Arbuscular Mycorrhizal Fungi-Associated Bacteria and Their Role in Plant Protection in Rice Cultivation

Mamun Mandal and Abhijit Sarkar
University of Gour Banga

CONTENTS

15.1 INTRODUCTION

One of humankind's significant difficulties will be producing sufficient food for an increasing worldwide population in the upcoming years. Reducing the use of artificial fertilizers and pesticides also preserve the natural resources and environmental quality for future generations. The new agricultural paradigm, sustainable development, can be applied through the efficient use of strategies that increase the beneficial soil microbes' activity. Soil microbes serve critical roles in human welfare and nutrition, and the economic value of ecosystem services they offer in forestry, agriculture, and society is becoming more widely recognized (Turrini et al., 2018). Rice (*Oryza sativa* L.) is considered one of the most important crops, feeding more than half of the entire world's population. Rice rhizosphere contains

DOI: 10.1201/9781003354086-15

a huge number of microbes that interact with rice roots and play a vital role in crop health management and long-term yield (Ding et al., 2019). Arbuscular mycorrhizal (AM) fungi (AMF) are one of the most significant rice plant symbionts, providing various soil nutrients in return for plant carbohydrates (Bao et al., 2022). Furthermore, AMF can boost rice plants' biotic and abiotic resilience. Additionally, the abundant extraradical hyphae formed by AMF in rice cultivated soil provide a residence for other microorganisms, including both rhizosphere and cytoplasmic microbes of some fungal species (Venice et al., 2020). However, there has been a lack of research on the symbiosis between AMF and microorganisms in the rice mycorrhizosphere. In this chapter, first we summarize the AMF-associated bacterial population of the rice mycorrhizosphere. Then, we discuss the mutualism processes involved between rice plants and AMF-related bacteria.

15.2 AMF-ASSOCIATED BACTERIAL COMMUNITY IN RICE MYCORRHIZOSPHERE

The zone impacted by both the root and the mycorrhizal fungus is characterized as the mycorrhizosphere. It is also described by the more precise word "hyphosphere," which refers specifically to the region around a particular hyphae of the fungus (Zhang et al., 2021). This limited zone is distinguished by enhanced microorganism activity caused through the exudation and leakage of organic compounds from the mycorrhiza and root that differ from the bulk soil (Frey-Klett et al., 2005). There are several kinds of microbial functional assembly that interact with AMF in the rice "mycorrhizosphere," which could be associated with variations in plants. Several multimodal studies revealed the presence of various assemblages of bacterial communities in the mycorrhizosphere that were firmly attached to AMF spores, mycorrhizal roots, and extraradical mycelium (Emmett et al., 2021). Previously, a study detect some free-living bacterial populations attached to the spore of AMF wall layers and the microniches developed by peridial hyphae interlinked surrounding the spores in the sporocarps (Turrini et al., 2018). AMF has been observed to interact with many kinds of soil bacteria such as plant growth promoting rhizobacteria (PGPR), mycorrhiza helper bacteria (MHB), endobacteria, and deleterious bacteria (DB). Therefore, validating such a concept and untangling the physiological interconnections between AMF and related bacteria may result in favorable additive impacts on plant nutrition and health.

15.2.1 Plant Growth Promoting Rhizobacteria

PGPR is essential to soil bacteria because it increases the rice output and minimize the requirement for artificial fertilizers (Hussain et al., 2021). The methods are as follows: (i) increasing soil nutrient solubility by synthesizing enzymes and siderophore production; (ii) phytohormones production; (iii) suppressing pathogens and mitigating the negative impacts of stress; and (iv) communicating with

TABLE 15.1
List of AMF-Associated Bacteria in Rice Mycorrhizosphere

AMF Species	Bacteria	Effects	References
Glomus sp., *Gigasporasp* sp., and *Acaulospora* sp.	*Azospirillum lipoferum* and *Bacillus megaterium* (MHB)	Increased rice plant growth and grain yield	Prem Kumari and Prabina (2017)
Glomus mosseae	*Burkholderia cepacia* and *Azospirillum brasilense*	Root biomass, root length, and crop yield by increasing phosphorus uptake in straw and grain	Beura et al. (2020)
G. mosseae	*Herbaspirillum seropedicae* (PGPR)	Increased rice growth and yield	Hoseinzade et al. (2016)
R. irregularis	*P. fluorescens* and *P. putida* (PGPR)	Increased rice grain yield under salinity stress conditions	Norouzinia et al. (2020)
F. mosseae	*Streptomyces thermocarboxydus* (Endobacteria)	Rice growth by producing siderophores, indole-3-acetic acid (IAA), and phosphate solubilization	Lasudee et al. (2018)
Rhizophagus intraradices	Actinobacteria, Acidobacteria, Bacteroidetes, Firmicutes, Chloroflexi, Gemmatimonadetes, Proteobacteria, Planctomycetes, and Verrucomicrobia	Reduce rice cadmium uptake by changing cadmium transporters expression and increased plant productivity	Chen et al. (2019)

certain other soil microorganisms (Bao et al., 2022). Table 15.1 represents the various kinds of AMF-associated PGPR and their activity in rice plant growth promotion.

15.2.2 Mycorrhiza Helper Bacteria

Garbaye (1994) was the first to identify MHB, which mostly contains Proteobacteria, Actinomycetes, Oxalobacteraceae, and Firmicutes (Bao et al., 2022). MHB can play a crucial role in boosting AMF growth and activity, giving nutrients to AMF and plants, and promoting their defenses (Gupta & Chakraborty, 2020). According to Rillig et al. (2005), MHB is generally not plant-specific but fungal-specific, which suggests that MHB can stimulate the establishment of a particular AMF even the associations of its nonspecific host plant. Furthermore, since there have been so few number studies conducted on MHB and AMF in rice, common unaddressed challenges (including whether the growth of AMF enhanced and survival through MHB are due to growth factors production, soil allelochemicals detoxification, and the antagonism of parasites and competitors) remain unsolved in rice (Frey-Klett et al., 2007).

15.2.3 Endobacteria

Some endobacteria within AMF cytoplasms are also known as PGPR, forming a close relationship with AMF. The Glomeromycota (example of endobacteria) are widely studied bacterial endosymbionts (Bonfante and Desirò, 2017). Glomeromycota has two types of endosymbionts: (i) Gram-negative, rod-shaped bacteria, beta-proteobacterium *Candidatus Glomeribacter gigasporarum* (Mondo et al., 2012) and (ii) Mollicutes-related endobacteria, coccoid bacteria with a homogenous gram-positive like cell wall structure that is found throughout the Glomeromycota (Naumann et al., 2010). The processes involved in endobacteria altering fungal performance such as chemicals released that alter fungal gene expression (such as volatiles), fungal cell wall destruction, the synthesis of lectins that adhere to the fungal surface, and the infusion of biomolecules into the fungal spores (Salvioli et al., 2016).

15.2.4 Deleterious Bacteria

AMF interact with DB and has a harmful effect on plant growth. Their negative impacts on plants are most likely related to the generation of toxic chemicals like phytotoxins and competition for nutrient sources with other soil microbes (Miransari, 2011). Several processes have been proposed to elucidate AMF antagonistic activity, including parasitizing pathogens, increased competition for colonization positions and host photosynthates, production of secondary metabolites, microbial community modification, plant nutrient uptake enhancement, root morphology modifications, and inducing plant systemic resistance (Bao et al., 2022). Although several harmful bacteria exist in the rice mycorrhizosphere, very few researches have been done in the influence of AMF on DB management. Contradictorily, PGPR can be harmful to mycorrhizal colonization, emphasizing the possibility of functional competition among helpful microbes (Hashem et al., 2016). This is determined by the characteristics of the mycorrhizosphere, like AM development and growth, microorganism growth phase, and environmental factors.

15.3 MECHANISMS INVOLVED IN RICE HEALTH MANAGEMENT BY AMF-ASSOCIATED BACTERIA

The combined application of AMF and other soil microorganisms helps plants, especially rice, to increase macro and micronutrient uptake (Bao et al., 2022). The importance of nutrient exchange in below-ground microbe and above-ground plant interactions is widely understood. Rice roots secrete carbon metabolites that promote rhizospheric microbial growth and functional activities (Hussain et al., 2021). By providing nutrients, phytohormones, and improving tolerance to biotic and abiotic stress, the microorganisms benefit the plants in return (Prem Kumari and Prabina, 2017). AMF interactions with microorganisms also involve nutrition exchanges in addition to those between AMF and rice plant mutualism

(Jamiołkowska et al., 2017). Through the broad extraradical hyphae adhering to the soil, AMF may deliver carbon molecules from plants to the soil microorganisms. During the same time, soil-dwelling bacteria return mineral nutrients to the AMF. The triangular relationship between AMF, rice plants, and microorganisms is necessary for rice field ecosystems to operate ecologically. Rice plants have various root exudates that attract and maintain microorganisms' diversity to forming mutualistic partnerships (Zhalnina et al., 2018). The substances are mostly amino acids, organic acids, hormones, polysaccharides, and other primary and secondary metabolites, all of which are derived from photosynthates (Matsushima et al., 2021). AMF-plant interactions also involve the AMF spores germination, the growth and differentiation of hyphal, the colonization of roots, and the promotion of the development of host-specific rhizobia by using flavonoids as chemoattractants (Mandal et al., 2010).

Some bacteria linked to AMF can produce several signaling chemicals, including volatile organic compounds, phytohormones, polysaccharides, jasmonic acid, abscisic acid, and ethylene synthesis inhibitors (rhizobitoxine, 1-aminocyclopropane-1-carboxylate deaminase), all of which are helpful for plant development and disease control (Khan et al., 2020). Some organisms can produce extracellular hydrolytic enzymes that can degrade a range of cell wall components of pathogens (Doni et al., 2019). Additionally, the development of associative symbiosis and colonization are connected by a variety of microbial-mediated activities, pathways (like quorum sensing), and signaling components. An essential characteristic of the plant-driven selection of microorganisms and subsequent colonization is microbial chemotaxis toward certain root-exuded chemicals. For PGPRs, the development of biofilms—a layer of bacterial cells associated with AMF, contained inside a self-produced exopolysaccharide matrix—determines their adhesion to AMF. Additionally, the bacteria can activate mitochondrial metabolic processes to promote AMF ecologically fitness, boost the fungal bioenergetic ability, enhance ATP synthesis, and elicit pathways to detoxifying reactive oxygen species (Ujvári et al., 2021). It's interesting to note that AMF lack the capacity to synthesize phosphatases, which prevent them from directly using organic nutrients. Instead, they mostly obtain their organic material through mineralization triggered by soil bacteria (Pandit et al., 2020). The chemicals that bacteria produce are essential for the efficient generation of microbial inoculums and for enhancing the bacteria's adhesion to AMF structures and mycorrhizal roots.

15.4 CONCLUDING OBSERVATIONS

For a long time, microbial consortia have drawn attention to their usefulness in the agricultural field. Currently, a commercial product with a patent called "Micosat F" has a combination of helper bacteria, AMF, and saprophytic fungi (Nuti & Giovannetti, 2015). A cornerstone of the upcoming "Green Revolution" in agriculture is the use of helpful bacteria in crop production. The communications between AMF and the bacteria in the rice rhizosphere are among the most essential and crucial factors affecting soil structure and rice development. For the triangular symbiotic relationship of rice plants, bacteria, and AMF, the nutritional exchange is crucial. This interaction can also be maintained by the substances exhaled by rice roots and microorganisms. Although certain bacterial consortia incorporating mycorrhizal inoculums are used in crop cultivation, our understanding of the complicated relationship between AMF and bacteria in the rice mycorrhizosphere is still in its infancy.

REFERENCES

Bao, X., Zou, J., Zhang, B., Wu, L., Yang, T., & Huang, Q. 2022. Arbuscular mycorrhizal fungi and microbes interaction in rice mycorrhizosphere. *Agronomy* 12(6):1277.

Beura, K., Pradhan, A. K., Ghosh, G. K., Kohli, A., & Singh, M. 2020. Root architecture, yield and phosphorus uptake by rice: Response to rock phosphate enriched compost and microbial inoculants. *Int. Res. J. Pure Appl. Chem.* 21:33–39.

Bonfante, P., & Desirò, A. 2017. Who lives in a fungus? The diversity, origins and functions of fungal endobacteria living in Mucoromycota. *ISME J.* 11:1727–1735.

Chen, X. W., Wu, L., Luo, N., Mo, C. H., Wong, M. H., & Li, H. 2019. Arbuscular mycorrhizal fungi and the associated bacterial community influence the uptake of cadmium in rice. *Geoderma* 337:749–757.

Ding, L. J., Cui, H. L., Nie, S. A., Long, X. E., Duan, G. L., & Zhu, Y. G. 2019. Microbiomes inhabiting rice roots and rhizosphere. *FEMS Microbiol. Ecol.* 95:fiz040.

Doni, F., Mispan, M. S., Suhaimi, N. S. M., Ishak, N., & Uphoff, N. 2019. Roles of microbes in supporting sustainable rice production using the system of rice intensification. *Appl. Microbiol. Biot.* 103:5131–5142.

Emmett, B. D., Lévesque-Tremblay, V., & Harrison, M. J. 2021. Conserved and reproducible bacterial communities associate with extraradical hyphae of arbuscular mycorrhizal fungi. *ISME J.* 15(8):2276–2288.

Frey-Klett, P., Chavatte, M., Clausse, M. L., Courrier, S., Le Roux, C., Raaijmakers, J., Martinotti, M. G., Pierrat, J. C., & Garbaye, J. 2005. Ectomycorrhizal symbiosis affects functional diversity of rhizosphere fluorescent pseudomonads. *New Phytol.* 165:317–328.

Frey-Klett, P., Garbaye, J., & Tarkka, M. 2007. The mycorrhiza helper bacteria revisited. *New Phytol.* 176:22–36.

Garbaye J. 1994. Tansley Review No. 76 Helper bacteria: a new dimension to the mycorrhizal symbiosis. *New Phytol.* 128:197–210. doi: 10.1111/j.1469-8137.1994.tb04003.x. PMID: 33874371.

Gupta, S. K. & Chakraborty, A. P. 2020. Mycorrhiza helper bacteria: Future prospects. *Int. J. Res. Rev.* 7:387–391.

Hashem, A., Abd_Allah, E. F., Alqarawi, A. A., Al-Huqail, A. A., Wirth, S., & Egamberdieva, D. 2016. The interaction between arbuscular mycorrhizal fungi and endophytic bacteria enhances plant growth of Acacia gerrardii under salt stress. *Front. Microbiol.* 7:1089.

Hoseinzade, H., Ardakani, M. R., Shahdi, A., Rahmani, H. A., Noormohammadi, G., & Miransari, M. 2016. Rice (*Oryza sativa* L.) nutrient management using mycorrhizal fungi and endophytic *Herbaspirillum seropedicae*. J. Integr. Agr. 15:1385–1394.

Hussain, Q., Pan, G. X., Liu, Y. Z., Zhang, A., Li, L. Q., Zhang, X. H., & Jin, Z. J. 2021. Microbial community dynamics and function associated with rhizosphere over periods of rice growth. *Plant Soil Environ.* 58:55–61.

Jamiołkowska, A., Ksiezniak, A., Hetman, B., Kopacki, M., Skwaryło-Bednarz, B., Gałazka, A., & Thanoon, A. H. 2017. Interactions of arbuscular mycorrhizal fungi with plants and soil microflora. *Acta Sci. Pol. Hortorum Cultus* 16:89–95.

Khan, N., Bano, A., Ali, S., & Babar, M. A. 2020. Crosstalk amongst phytohormones from planta and PGPR under biotic and abiotic stresses. *Plant Growth Regul.* 90:189–203.

Lasudee, K., Tokuyama, S., Lumyong, S., & Pathom-Aree, W. 2018. Actinobacteria associated with arbuscular mycorrhizal *Funneliformis mosseae* spores, taxonomic characterization and their beneficial traits to plants: Evidence obtained from mung bean (*Vigna radiata*) and thai jasmine rice (*Oryza sativa*). *Front. Microbiol.* 9:1247.

Mandal, S. M., Chakraborty, D., & Dey, S. 2010. Phenolic acids act as signaling molecules in plant-microbe symbioses. *Plant Signal. Behav.* 5:359–368.

Matsushima, C., Shenton, M., Kitahara, A., Wasaki, J., Oikawa, A., Cheng, W. et al. 2021. Multiple analysis of root exudates and microbiome in rice (Oryza sativa) under low P conditions. *Arch. Microbiol.* 203:5599–5611.

Miransari, M. 2011. Interactions between arbuscular mycorrhizal fungi and soil bacteria. *Appl. Microbiol. Biot.* 89:917–930.

Mondo, S. J., Toomer, K. H., Morton, J. B., Lekberg, Y., & Pawlowska, T. E. 2012. Evolutionary stability in a 400-million-year-old heritable facultative mutualism. *Evolution* 66:2564–2574.

Naumann, M., Schüler, A., & Bonfante, P. 2010. The obligate endobacteria of arbuscular mycorrhizal fungi are ancient heritable components related to the Mollicutes. *ISME J.* 4:862–871.

Norouzinia, F., Ansari, M. H., Aminpanah, H., & Firozi, S. 2020. Alleviation of soil salinity on physiological and agronomic traits of rice cultivars using Arbuscular mycorrhizal fungi and Pseudomonas strains under field conditions. *J. Neotrop. Agri.* 7:25–42.

Nuti, M. & Giovannetti, G. 2015. Borderline products between bio-fertilizers/bio-effectors and plant protectants: The role of microbial consortia. *J. Agric. Sci. Technol.* 5:305–315.

Pandit, A., Adholeya, A., Cahill, D., Brau, L., & Kochar, M. 2020. Microbial biofilms in nature: Unlocking their potential for agricultural applications. *J. Appl. Microbiol.* 129:199–211.

Prem Kumari, S. M. & Prabina, B. J. 2017. Impact of the mixed consortium of indigenous arbuscular mycorrhizal fungi (AMF) on the growth and yield of rice (*Oryza Sativa* L.) under the system of rice intensification (SRI). Inter. J. Environ. Agri. Biotechnol. 2:238743.

Rillig, M. C., Lutgen, E. R., Ramsey, P. W., Klironomos, J. N., & Gannon, J. E. 2005. Microbiota accompanying different arbuscular mycorrhizal fungal isolates influence soil aggregation. *Pedobiologia* 49:251–259.

Salvioli, A., Ghignone, S., Novero, M., Navazio, L., Venice, F., Bagnaresi, P., & Bonfante, P. 2016. Symbiosis with an endobacterium increases the fitness of a mycorrhizal fungus, raising its bioenergetic potential. *ISME J.* 10:130–144.

Turrini, A., Avio, L., Giovannetti, M., & Agnolucci, M. 2018. Functional complementarity of arbuscular mycorrhizal fungi and associated microbiota: The challenge of translational research. *Front. Plant Sci.* 9:1407.

Ujvári, G., Turrini, A., Avio, L., & Agnolucci, M. 2021. Possible role of arbuscular mycorrhizal fungi and associated bacteria in the recruitment of endophytic bacterial communities by plant roots. *Mycorrhiza* 31(5):527–544.

Venice, F., Ghignone, S., Salvioli di Fossalunga, A., Amselem, J., Novero, M., Xianan, X. et al. 2020. At the nexus of three kingdoms: The genome of the mycorrhizal fungus

Gigaspora margarita provides insights into plant, endobacterial and fungal interactions. Environ. Microbiol. 22:122–141.
Zhalnina, K., Louie, K. B., Hao, Z., Mansoori, N., da Rocha, U. N., Shi, S. et al. 2018. Dynamic root exudate chemistry and microbial substrate preferences drive patterns in rhizosphere microbial community assembly. *Nat. Microbiol.* 3:470–480.
Zhang, L., hou, J., George, T. S., Limpens, E., & Feng, G. 2021. Arbuscular mycorrhizal fungi conducting the hyphosphere bacterial orchestra. *Trends Plant Sci.* 27:402–411.

16 Arbuscular Mycorrhizal Fungi and Their Association for Bioremediation in Rice Cultivation

Shuvendu Shekhar Mohapatra
ICAR-National Rice Research Institute
Berhampur University

Anuprita Ray
Lovely Professional University

Sonali Panda
ICAR-National Rice Research Institute
Ravenshaw University

Sucharita Satapathy
ICAR-National Rice Research Institute

Nutan Moharana
ICAR-National Rice Research Institute
Odisha University of Agriculture and Technology

CONTENTS

DOI: 10.1201/9781003354086-16

16.1 INTRODUCTION

The process of bioremediation has incorporated managing the polluted or contaminated ecosystems where several microorganisms are engaged to accelerate the natural procedures that detoxicate/decontaminate the soil ecosystem. The possible role of remediation by microorganisms, predominantly higher terrestrial crops/plants (phytoremediation) explored in the decontamination of heavy metal-polluted zones, has been the prime focus of significant research in the modern era. AMF are the microorganisms in the soil that set up mutual symbiosis (root colonization) with the common cereal plants that participated in phytoremediation that consider crops for soil remediation (Verma et al., 2019).

Bioremediation is the use of several specific organisms for the handling of polluted soil. AMF are root-colonizing symbiotic microorganisms mainly involved in phytoremediation where plants are used for soil reclamation. The process of phytoremediation encompasses a set of protocols that use several plants as a control, destruction, or extraction performance. Compared to conventional nonbiological methods, these techniques have gained interest owing to their influential cost savings. Various techniques of phytoremediation may be useful depending on the type of pollutants. Phyto-extraction, the extraction of heavy metals from the soil, is the only method by which heavy metals can be removed from soil or restrained in a nontoxic form (phyto-stabilization). AMF can lessen heavy metal toxicity in plants by dropping metal translocation from the root system to the shoot (Mishra et al., 2019). As a result, they add to plant establishment and endurance in heavy metal-polluted lands and could be cast off as a complement to hold strategies. Phyto-extraction primarily uses plants accumulating elevated concentrations of heavy metals, which can be harvested, discarded, and even extracted to recover metals. This is why various plants are used, including *Brassicaceae* members, which are typically thought to not be mycorrhizal, as well as other accumulators that produce higher biomass and may be mycotrophic (Yao et al., 2012). The microbial action, which is frequently boosted in the root colonies, can change or break down organic contaminants like polycyclic aromatic hydrocarbons (PAH) (rhizo-degradation), and it is unclear if the increased microbial activity or plant exudates, which may contain several enzymes, surfactants, and various physical/chemical effects, are to blame for the rhizosphere's heightened degradation. Another potential method for the oxidation of organic contaminants might be reliant on the metabolic processes of plants. But this route for PAH is not particularly significant in terms of quantity. Because AMF alter root transudation and root-associated populations of microbes, and in certain ways operate as a postponement of the roots' external binding of the rhizosphere, it may be advantageous for PAH rhizo-degradation. This may also have a direct consequence on PAH degradation. Eventually, AMF can be cast off in bio-linkage of soil quality or soil toxicity, in arrears to their compassion toward a bunch of soil pollutants (Igiri et al., 2018). Enough soil quality needs to be reestablished after bioremediation by using AMF. This chapter assesses the potential of AMF in heavy metal-polluted soils, their role in conferring metal tolerance to plants, the factors affecting AMF in metal-polluted soils, and their driving forces toward heavy metal toxicity tolerance.

16.2 AMF-MEDIATED PHYTOREMEDIATION OF HEAVY METAL-POLLUTED SOILS

The accessibility of metal cations, to plants, mainly depends on their mobilization and immobilization by soil microorganisms. AMF make up a significant functional constituent of the soil-plant structure and are among the most common soil microorganisms stirring in several habitats and climates, including concerned soils. Degraded land soils do, nevertheless, agonize from variations in diversity and plenty of AMF diversity (Mishra et al., 2019). More precisely, it has been reported that AMF can be exaggerated by heavy metal stress, but in a few cases, mycotrophic crops growing in land soils polluted with heavy metals are colonized by AMF (Verma et al., 2019). In the previous few years, research attention has been advanced on the varietal improvement and tolerance of AMF in heavy metal-polluted soils annoying to comprehend the elementary underlying variation and tolerance of AMF to heavy metal toxicity in soils; meanwhile, this could ease the supervision of these soil microbes, for rebuilding bioremediation approaches.

16.3 HEAVY METAL TOXICITY AND RICE SOIL

There are several heavy metals present in surface soil, and the elevated amount of most of them causes much damage to the growth and regulation of plants. Heavy metals like iron, arsenic, lead, copper, and many others have toxic effects on agricultural practices and their traces in food cause several health issues (Jena et al., 2018). All aspects of heavy metal toxicity and their remediation through AMF are discussed below (Mitra et al., 2022).

16.3.1 Effects of Arsenic Toxicity on Soil and Rice Plant

Rice (*Oryza sativa* L) is cultivated in a semi-aquatic mode and is the only prime source of all bioavailable nutritional requirements of about half of the global population. India is the second largest rice producer in the world after China and exports to all other major developed countries globally. Though more than 40,000 rice varieties are developed worldwide, few of them are cultivated extensively and are adopted by most farmers. In this current scenario, studies have reported higher levels of arsenic (As) in soil ecosystems and rice is a major concern to be considered. Among the most hazardous substances, arsenic is the most harmful metal having characteristics of carcinogenic and toxicity. In Asian countries like China,

India, and Bangladesh, where rice is cultivated spontaneously, fields are exposed majorly to arsenic contamination associated with inappropriate irrigation and lower levels of groundwater. Several ways through which arsenic enters the soil ecosystem are the application of chemicals like pesticides and insecticides, which are generally a mixture of several metallic traces like arsenic to the agricultural land. For a long time, higher levels of arsenic have been present in the soil, and groundwater is now prone to rice grains and may cause several hazards to millions of rice-consuming people globally. The ever-increasing mixing of arsenic residues in herbicides, pesticides, paints, textile industries, and several other household products increases the concentration of arsenic in the soil and groundwater. A statement from the Indian Council of Agricultural Research (ICAR) suggests weed-killing herbicides and insecticides are having arsenic compounds in higher amounts. Insecticides, for example, calcium arsenate ($Ca_3As_2O_8$), lead arsenate ($PbHAsO_4$) with generally applied herbicides such as sodium arsenite ($NaAsO_2$) as well as rodenticides, including arsenic trioxide (As_2O_3), applied primarily for root diseases. When all these pesticides are sprayed on agricultural crops or soils, their traces remain in the soil for a long; traces are mixed in groundwater and are offered to living microorganisms for remediation prospects (Kalita et al., 2018). It has been proved that the inorganic states of arsenic are deposited in soil for more than decades, their oxidized form is arsenate (As^{5+}), and is generally present in oxygen-enriched soil. While less oxygen is present in soil, arsenic prevails to be arsenite (As^{3+}) (Awasthi et al., 2017; Kalita et al., 2018).

16.3.1.1 Arsenic Toxicity and Its Effect on Humans

Captivatingly, rice can accumulate arsenic traces fifteen times more than other cereal crops like rice and wheat. Arsenic (As) can aggravate some degenerative aspects of several metabolic processes, development of crops, growth in the vegetative stage, and yield parameters of the plants. Many studies report that the mechanism of the arsenite transport system in rice is primarily carried out by silicon (Si) transporter OsNIP2; 1 (OsLsi1), and the downflow from roots to xylem membrane is started by OsNIP2; 1 (OsLsi2) (Ma et al., 2008). The presence of an elevated amount of arsenic traces in rice plants decreases the percentage of photosynthesis and chlorophyll content, prompting a lessening in the crop biomass production and factors affecting grain yield characters. In humans, arsenic-facilitated toxicity has a negative impact on health, like dysfunction of the liver, respiratory tract disease, nervous and brain disorders, gastrointestinal problems, cardiac diseases, and circulatory diseases like anemia and diabetes. The past drawbacks of Mandal and Suzuki (2002) stated that 80% of arsenic uptake in human beings is due to this intoxicated food consumption (rice and maze), and direct inhalation counts for only $<10\%$ (Mandal and Suzuki, 2002).

16.3.1.2 AMF-Mediated Bioremediation of Arsenic

AMF help in the toxicity remediation of several heavy metals by the phylum Glomeromycota and count for about 80% of almost all plant species (Mitra et al., 2021). These types of obligate symbiotic fungi were sorted over millions of years, and this cluster of microorganisms can form colonization in the roots

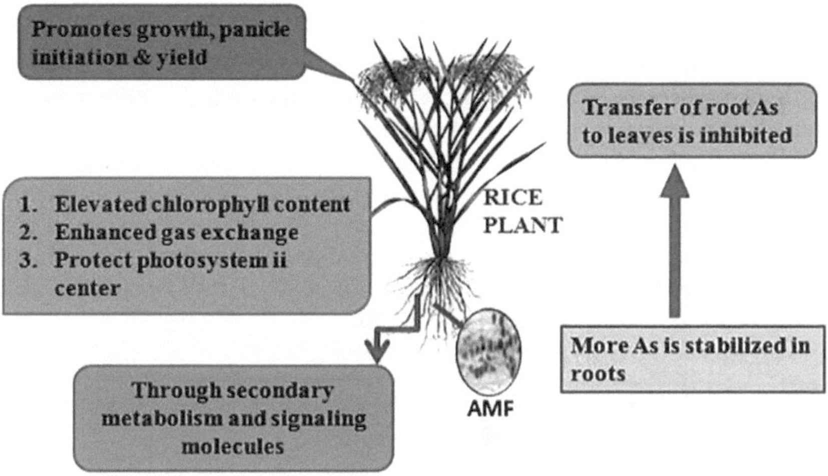

FIGURE 16.1 Association of arbuscular mycorrhizal fungi in rice plant root system and the diagrammatic representation of various biochemical and physiological aspects under arsenic toxicity.

of the common land crops in the specificity of host cell interactions. The profusion and AMF communities in the rhizosphere of diverse vascular plants show the prevalence of these microorganisms in bearable agricultural systems (Gómez Expósito et al., 2017). The major favorable influences of Indian-originated AMF are an enlargement in the host crop's water and mineral uptake processes and correlate with another physio-chemical process of plants like rice (Figure 16.1). The loads and groups of AMF present in rhizosphere nodes have two elements that manipulate nutritional replenishment in plants and prolonged crosstalk in plants (Prasad et al., 2005). The mutual relationship represents an improvement of the host plant to perk the adaptation toward several biotic and metallic stress management. AMF have an extensive past of co-evolution with the host plant in several habitats ensuing in various regions (Gosling et al., 2006).

16.3.2 Lead (Pb) Toxicity and Arbuscular Mycorrhizal Fungi

Lead (Pb) is a heavy metal that causes toxicity and is present in fewer concentrations on agricultural lands unless it is affected by industrial pollutants. Anthropogenic activities may increase its concentration in several places, which could lead to unproductive or low yield of several crops. Under the lead (Pb) toxicity/stress, the unfavorable oxidative properties adversely affect the growth of several plants or crops in general. However, AMF have the potential to create tolerance and increase the growth and productivity in several agricultural practices. It can enhance the antioxidant enzymes, which can elevate the stress of lead (Pb) (Dash and Mohapatra, 2018).

16.3.3 Iron (Fe) Concentration and Arbuscular Mycorrhizal Fungi

A decrease in the concentration of iron (Fe) in agricultural land practices is one of the prime constraints, which can affect plant growth, yield, and other nutritional parameters in several kinds of cereal. Studies have been performed to understand and establish the process of how AMF uplift the Fe-concentration deficiency in several crops (Gosling et al., 2006). The inoculation of AMF improves plant biomass development, chlorophyll score (quantum efficiency of photosystem II and photosynthesis performance index), and reduction in cell death. Overall, AMF enhance and upregulate the growth of plants through root colonization and other biochemical processes (Rahman et al., 2020).

16.3.4 Copper (Cu) Toxicity and Arbuscular Mycorrhizal Fungi

Elevated levels of copper (Cu) in soil may cause toxicity in young seedlings or plants by reducing the absorption capability of roots. Studies have reported that the inoculation of AMF in cropping systems like rice or wheat enhances the ability to absorb water and minerals by symbiosis in root colonies (Ambrosini et al., 2015). It directly enhances plant growth and better photosynthesis by enhancing chlorophyll content. It also improves grain quality and other nutritional parameters by regulating biochemical events that occur naturally without any stress that is eliminated or reduced by AMF (Ambrosini et al., 2015).

16.4 CONCLUSION AND FUTURE PROSPECTIVE

Contamination of water and soil is currently a major concern worldwide. Heavy metal toxicity in soil water and its adverse effects on humans are very severe and its bioremediation is enormously felt, for which the understanding of symbiosis-like mycorrhizal fungi association and its exceptional effects on heavy metals like arsenic (As), iron (Fe), lead (Pb), copper (Cu), and others accepted by crops is appealing. AMF application not just decreases this heavy metal toxicity but also regulates plant growth and development, chlorophyll content, and nutrient uptake by plants. AMF have an important influence on the interactions of both organic and inorganic materials. The significant impact of AMF in falling heavy metal uptake in the shoot, root, and reshuffle to grains in genotypes has vital suggestions for delivering toxin-free grains to heavy metal-affected.

REFERENCES

Ambrosini, V.G., Voges, J.G., Canton, L., Couto, R.D.R., Ferreira, P.A.A., Comin, J.J., Melo, G.W.B.D., Brunetto, G. and Soares, C.R.F.S., 2015. Effect of arbuscular mycorrhizal fungi on young vines in copper-contaminated soil. Brazilian Journal of Microbiology, *46*, pp. 1045–1052.

Awasthi, S., Chauhan, R., Srivastava, S. and Tripathi, R.D., 2017. The journey of arsenic from soil to grain in rice. Frontiers in Plant Science, *8*, p. 1007.

Dash, A. and Mohapatra, S.S., 2018. Toxic effect of urea on earthworms determined by a simple paper contact method. *Innovare Journal of Agricultural Sciences, 6*(1). pp. 2321–6832.

Gómez Expósito, R., De Bruijn, I., Postma, J. and Raaijmakers, J.M., 2017. Current insights into the role of rhizosphere bacteria in disease suppressive soils. Frontiers in Microbiology, *8*, p. 2529.

Gosling, P., Hodge, A., Goodlass, G. and Bending, G.D., 2006. Arbuscular mycorrhizal fungi and organic farming. *Agriculture, Ecosystems & Environment, 113*(1–4), pp. 17–35.

Igiri, B.E., Okoduwa, S.I., Idoko, G.O., Akabuogu, E.P., Adeyi, A.O. and Ejiogu, I.K., 2018. Toxicity and bioremediation of heavy metals contaminated ecosystem from tannery wastewater: a review. Journal of Toxicology, *2018*, p. 2568038.

Kalita, J., Bairagi, P., Kashyap, G., Sarma, M.K., Baruah, S. and Sharma, A.A., 2018. Genetic variability and character association studies in indigenous ahu rice germplasm of Assam, India. *International Journal of Current Microbiology and Applied Sciences, 7*(6), pp. 2297–2304.

Ma, J.F., Yamaji, N., Mitani, N., Xu, X.Y., Su, Y.H., McGrath, S.P. and Zhao, F.J., 2008. Transporters of arsenite in rice and their role in arsenic accumulation in rice grain. *Proceedings of the National Academy of Sciences, 105*(29), pp. 9931–9935.

Mandal, B.K. and Suzuki, K.T., 2002. Arsenic round the world: a review. *Talanta, 58*(1), pp. 201–235.

Mishra, A., Bhattacharya, A., and Mishra, N. 2019. Mycorrhizal symbiosis: An effective tool for metal bioremediation. In *New and Future Developments in Microbial Biotechnology and Bioengineering*; Singh, J.S., Ed.; Elsevier: Amsterdam, The Netherlands, pp. 113–128.

Mitra, D., Djebaili, R., Pellegrini, M., Mahakur, B., Sarker, A., Chaudhary, P., Khoshru, B., Gallo, M.D., Kitouni, M., Barik, D.P. and Panneerselvam, P., 2021. Arbuscular mycorrhizal symbiosis: plant growth improvement and induction of resistance under stressful conditions. Journal of Plant Nutrition, *44*(13), pp. 1993–2028.

Mitra, D., Saritha, B., Janeeshma, E., Gusain, P., Khoshru, B., Nouh, F.A.A., Rani, A., Olatunbosun, A.N., Ruparelia, J., Rabari, A. and Mosquera-Sánchez, L.P., 2022. Arbuscular mycorrhizal fungal association boosted the arsenic resistance in crops with special responsiveness to rice plant. Environmental and Experimental Botany, *193*, p. 104681.

Prasad, R., Pham, H.G., Kumari, R., Singh, A., Yadav, V., Sachdev, M., Garg, A.P., Peskan, T., Hehl, S., Sherameti, I. and Oelmuller, R., 2005. Sebacinaceae: culturable mycorrhiza-like endosymbiotic fungi and their interaction with non-transformed and transformed roots. In In Vitro Culture of Mycorrhizas; Declerck, S., André Fortin, J., and Strullu, D.-G., Eds.; Springer: Berlin, Heidelberg, pp. 291–312.

Rahman, M.A., Parvin, M., Das, U., Ela, E.J., Lee, S.H., Lee, K.W. and Kabir, A.H., 2020. Arbuscular mycorrhizal symbiosis mitigates iron (Fe)-deficiency retardation in alfalfa (*Medicago sativa* L.) through the enhancement of Fe accumulation and sulfur-assisted antioxidant defense. *International Journal of M*olecular Sciences, *21*(6), p. 2219.

Verma, R., Tapwal, A., Kumar, D., Parkash, V. and Puri, S., 2019. Vesicular arbuscular mycorrhizal diversity in some important ethnomedicinal plants of Western Himalaya. *Medicinal Plants, 11*, pp. 279–285.

Yao, Z., Li, J., Xie, H. and Yu, C., 2012. Review on remediation technologies of soil contaminated by heavy metals. Procedia Environmental Sciences, *16*, pp. 722–729.

17 AM Fungi Interactions in Rice Seedling Production

Nurudeen Olatunbosun Adeyemi and Oni Olanrewaju Emmanuel
Federal University of Agriculture

Debasis Mitra
Raiganj University

CONTENTS

17.1 INTRODUCTION

Arbuscular mycorrhizal fungi (AMF) are members of the phylum Glomeromycota and form symbiotic relationships with the roots of most terrestrial plant species, including many important crops like rice (*Oryza sativa* L). They play important ecological roles in the transport of water and essential nutrients such as P and N, as well as the absorption of other micronutrients such as Cu, Zn, and Fe (Panneerselvam et al., 2013; Adeyemi et al., 2021a,b), in exchange for 20%–30% of photosynthetically fixed carbon (C) from the host plants (Parniske, 2008). They also play an important function in stress reduction and bioremediation by immobilizing heavy metals and preventing their transfer to plant aerial parts (Adeyemi et al., 2021a–c). There is also an ecological relevance in improving soil quality and protecting plants from pathogens. They can preserve the profitability and sustainability of agricultural systems. However, the prospect of AMF interaction with rice, particularly at the nursery under various growing circumstances, is poorly investigated.

DOI: 10.1201/9781003354086-17

Rice is an essential agricultural crop for more than half of the world's population. It is also used as a monocotyledon model plant for research on AM mutualistic relationships with plants (Sawers et al., 2018; Wang et al., 2021; Panneerselvam et al., 2017). Understanding the interaction of rice plants with AM fungi will assist in sustainable rice production from the nursery stage through the harvest.

17.2 ARBUSCULAR MYCORRHIZAL FUNGI SYMBIOSIS IN RICE PLANTS

A range of biotic and abiotic factors influence the interaction of rice plants with AMF, including cultivation conditions, rice genotype, AMF species, irrigation regimes, fertilizer input, diseases, and insect pests (Cosme et al., 2011). Cultivation conditions, variety, and cultural practices are crucial driving factors that impact rice plant productivity in response to AMF symbiosis. Inundation, in particular, has a substantial impact on root architecture and morphology, as well as soil parameters such as redox potential and pH (Wang et al., 2016; Ding et al., 2019). Flooding can also impact the photosynthetic process and, as a result, the C level of rice and other plants (Zhang et al., 2017), which is another important factor that may influence rice mycorrhizal performance in paddy fields (Konvalinková et al., 2015).

Shaded rice seedlings have low plant C levels, which limit AMF symbiosis owing to increased C competition between plants and AMF (Konvalinková and Jansa, 2016). Shading of rice seedlings in the nursery can also affect MGR by influencing P uptake in host plants and even brief periods of shade can have a significant impact on the functioning of the mycorrhizal symbiosis in terms of P transfer between the fungus and the plants (Konvalinková et al., 2015). To assess the possible consequences of AMF symbiosis in rice seedlings, various additional host plant characteristics must be considered. Increased rainfall and decreased hours of sunlight during the rice growing period are common in rice-producing areas because of global climate change (Ray et al., 2019), and thick cloud and particle pollution may further reduce solar irradiance (Tie et al., 2016). Remarkably, although the rice mycorrhizal response to different flooding regimes has been thoroughly described (Bao et al., 2019), the influence of plant C levels on rice mycorrhizal responses to flooding has yet to be published. Surveying of AMF and rice interaction in wetland habitats is vital for understanding their biological roles as well as the AMF application potential against the backdrop of a growing global population and increased environmental pollutants.

According to several studies, AMF are sparse or absent in rice roots in flooded paddy fields. Conversely, several studies have shown AMF colonization inside rice roots in paddy fields (Bernaola et al., 2018), despite floods having a suppressive impact on AMF colonization in rice roots (Zhang et al., 2015). Previous research discovered that AMF colonization in rice roots was prevalent during the heading and ripening stages, but absent or rare during the early growth stages (Wang et al., 2015), which could be explained in part by the well-developed aerenchyma, which allows atmospheric oxygen to be delivered to rice roots during the heading and ripening stages (Wang et al., 2015). Some studies have also looked at the effect of AMF inoculation on rice plant development in flooding (Bernaola et al., 2018) and non-flooding circumstances (Zhang et al., 2015). Rice plants' mycorrhizal growth response (MGR) ranged from positive to negative. Under flooding circumstances, most studies found that AMF boosted rice plant biomass, grain production, and P absorption (Gewaily, 2019). Others, on the contrary, discovered that AMF inoculation reduced the quantity of rice dry matter under flooded circumstances (Cosme et al., 2011; Bao et al., 2019). These findings add to our understanding of the ecological roles of AMF in paddy wetlands. However, their inconsistency in results regarding rice plant MGR, which should be closely related to the complex environment of paddy fields and the different experimental conditions of the studies, threatens a comprehensive understanding of the application prospects of AMF in rice-production systems (Bao et al., 2019). They also emphasized the need to take major influencing factors into account when evaluating the functional significance of AMF in paddy soils.

17.3 ROLE OF AM FUNGI FOR SUSTAINABLE MANAGEMENT IN RICE SEEDLING PRODUCTION

Although AMF are significant components of paddy fields and other wetlands (Wang et al., 2015), their functional significance in paddy wetlands is not entirely understood (Bao et al., 2019). Rice plants readily develop mycorrhizal associations in upland conditions, while the infection is uncommon in submerged conditions due to the frequency of anoxic environments (Wang et al., 2015). AMF symbioses with rice plant roots significantly boost insoluble P mobilization and absorption by plants. AMF-inoculated rice plants improve P absorption from weakly soluble P via direct or indirect mechanisms (Panneerselvam et al., 2016; Adeyemi et al., 2019, 2020). AMF may be utilized to boost plant development by increasing the availability of plant nutrients with poor mobility in soil, such as P, Zn, and Cu (Adeyemi et al., 2021a,b). Because of the saprophytic character of soil and the changing of numerous physical qualities such as water-holding capacity and porosity, soil organic matter content can influence the formation and activity of AMF. AMF-rice root association might boost the availability of both natural and added Zn in rice. AMF absorb and transport N, P, K, Ca, S, Fe, Cu, Mn, and Zn from the soil to the host plant. Aerobic rice genotypes treated with mycorrhizal inoculum demonstrated a 28%–57% increase in root colonization (Gao et al., 2007; Mitra et al., 2020). In the instance of upland rice seedlings, colonization with arbuscular mycorrhiza resulted in greater yields at optimal fertilizer dosage, lowering costs, as well as environmental damage. Upland ecology's natural flora of AMF has been proven to be quite efficient, and upland rice is typically sensitive to AMF (Panneerselvam et al., 2017, 2020; Maiti et al., 2011, 2012).

Drought reduces the output to some extent in around half of the world's rice acreage because there is insufficient water to sustain flooded conditions. Water stress during crucial phases might result in a significant reduction in output and crop loss. As a result, water management is required for sustainable rice cultivation. Increased plant water uptake, which may be done by inoculating AMF symbiotically connected with plant roots and therefore enhancing crop yield and plant development under stressed conditions (Adeyemi et al., 2021a–c), is one option for maintaining this sustainability. Rice plants commonly develop mycorrhizal associations in upland environments, whereas infection is extremely rare in submerged situations due to anoxic circumstances. Rice root colonization by AMF is stimulated by non-flooded rice farming and prevailing aerobic environment in soil (Vallino et al., 2014).

AMF rice has a high amount of glutathione when well-watered. However, under drought stress conditions, AMF rice has a lower amount of glutathione. The reduction in glutathione concentration in AMF plants during drought conditions increases lipid peroxidation (Ruíz-Sánchez et al., 2011). Additionally, AMF inoculation boosted ascorbic and proline levels, which are particularly efficient protective chemicals against the adverse effects of water scarcity. As a result, AMF are useful for rice plants as an environmentally clean and acceptable method for improving plant performance and growth. These findings indicate that the flooding regime is critical in defining the functional role of AMF in rice farming systems. AMF-rice interaction may be considerably influenced by factors such as AMF species identification (Chan et al., 2013), rice types (Huang et al., 2020; Panneerselvam et al., 2019), and soil conditions (Fiorilli et al., 2015). AMF colonization reduced the amount of shoot dry matter under flooded settings while increasing the output of unhulled grains under non-flooded conditions

Furthermore, C-P exchange between host plants and AMF is essential for AM symbiosis and has been shown to be a critical component in predicting symbiosis success (Corrêa et al., 2014). AMF are fully reliant on photosynthetically fixed C from their hosts to build intraradical and extraradical structures (Bravo et al., 2016). Previous research has found that flooding reduces C allocation from host plants to AMF (Bao et al., 2019; Panneerselvam et al., 2017).

17.4 AMF-INDUCED RESISTANCE IN RICE

AMF can improve plants' biotic resistance to soilborne fungal and bacterial diseases, nematodes, or root-chewing insects, as well as their tolerance to abiotic stress, such as shading stress, drought, excessive salinity, and heavy metal pollution (Campo et al., 2020; Adeyemi et al., 2021a,b). This has been claimed to be accomplished in part by inducing basal resistance in host plants (Cornejo et al., 2017). In rice, the activation of the defensive response to pathogen infection is indicated by the induction of pathogenesis-related (PR) genes (Mitsuhara et al., 2008), as well as the colonization of AMF or other mutualists (Tian et al., 2019). Rice plants can fine-tune their responses to balance growth and defense by modulating plant hormones (Zipfel and Oldroyd, 2017). Given the necessity of generating disease-resistant crops and minimizing fertilizer consumption in the current agricultural production, plant breeders must understand nutrient absorption and disease resistance.

Plant basal resistance mechanisms serve critical functions in regulating plant compatibility with AMF (Wang et al., 2021). AMF have gained the ability to bypass defensive systems controlled by the plant's immune system (Wang et al., 2021). Nonetheless, host plants display certain defense responses after the first AMF interaction (Campos-Soriano et al., 2012), which are turned off after a few days to allow for repeated colonization in a suitable AMF host. This AMF-triggered defense response, known as "defense priming," can lead to more efficient activation of defense systems in response to future attackers' attacks (Fiorilli et al., 2018). This AMF-induced defensive response may improve rice plants' ability to fight illness. This implies that the diverse functions of AM symbiosis and their consequences on rice performance should be investigated from multiple viewpoints in different situations. Two recent investigations (Fiorilli et al., 2015) found varied mycorrhizal responses across various root types in rice and indicated AMF-induced defensive responses in fine lateral roots, which are less compatible with AMF colonization. In this study, shading-induced carbon deficiency in rice plants greatly boosted the potency of the AMF-triggered defensive responses. The enhanced defensive response to the fungal partner might explain some of the decreased AMF colonization and MPR found in this and previous investigations (Konvalinková et al., 2015; Konvalinková and Jansa, 2016). These findings suggest that AM symbiosis in rice plants is fine-tuned in response to changing environmental conditions. In other words, rice roots carefully coordinate their defensive responses to adjust the carbon-consuming AMF based on the richness of carbon sources (Neuenkamp et al., 2021). More research into this process is still needed.

17.5 CONCLUSION

Understanding the role of the AM fungus in rice seedlings has significant implications for the use of AMF in rice-production systems. More research is needed to uncover the processes involved in AMF interactions with rice plants under diverse conditions, as well as in the regulation of mycorrhizal association in rice plants.

REFERENCES

Adeyemi, N.O., Atayese, M.O., and Olubode, A.A. (2019). Identification and relative abundance of native arbuscular mycorrhizal fungi associated with oil-seed crops and maize (*Zea mays* L.) in derived savannah of Nigeria. *Acta fytotechn zootechn* 22(3), 84–89.

Adeyemi, N.O., Atayese, M.O., Olubode, A.A., and Akan, M.E. (2020). Effect of commercial arbuscular mycorrhizal fungi inoculant on growth and yield of soybean under controlled and natural field conditions. *J. Plant Nutr.* 43(4), 487–499. doi: 10.1080/01904167.2019.1685101.

Adeyemi, N.O., Atayese, M.O., Sakariyawo, O.S., Azeez, J.O., Olubode, A.A., Ridwan, M., Adebiyi, A., Oni, O., and Ibrahim, I. (2021a). Influence of different arbuscular mycorrhizal fungi isolates in enhancing growth, phosphorus uptake and grain yield of soybean in a phosphorus deficient soil under field conditions. *Commun. Soil Sci. Plant Anal.* 52(10), 1171–1183.

Adeyemi, N.O., Atayese, M.O., Sakariyawo, O.S., Azeez, J.O., and Ridwan, M. (2021b). Arbuscular mycorrhizal fungi species differentially regulate plant growth, phosphorus uptake and stress tolerance of soybean in lead contaminated soil. *J. Plant Nutr.* 44(11), 1633–1648.

Adeyemi, N.O., Atayese, M.O., Sakariyawo, O.S., Azeez, J.O., Sobowale, S.P.A., Olubode, A., Mudathir, R., Adebayo, R., and Adeoye, S. (2021c). Alleviation of heavy metal stress by arbuscular mycorrhizal symbiosis in *Glycine max* (L.) grown in copper, lead and zinc contaminated soils. *Rhizosphere* 18, 100325.

Bao, X.Z., Wang, Y.T., and Olsson, P.A. (2019). Arbuscular mycorrhiza under water-carbon-phosphorus exchange between rice and arbuscular mycorrhizal fungi under different flooding regimes. *Soil Biol. Biochem.* 129, 169–177. doi: 10.1016/j.soilbio.2018.11.020.

Bernaola, L., Cange, G., Way, M.O., Gore, J., Hardke, J., and Stout, M. (2018). Natural colonization of rice by arbuscular mycorrhizal fungi in different production areas. *Rice Sci.* 25, 169–174. doi: 10.1016/j.rsci.2018.02.006.

Bravo, A., Brands, M., Wewer, V., Dörmann, P., and Harrison, M.J. (2016). Arbuscular mycorrhiza-specific enzymes FatM and RAM2 fine-tune lipid biosynthesis to promote development of arbuscular mycorrhiza. *New Phytol.* 214, 1631–1645. doi: 10.1111/nph.14533.

Campo, S., Martín-Cardoso, H., Olivé, M., Pla, E., Catala-Forner, M., Martínez Eixarch, M., et al. (2020). Effect of root colonization by arbuscular mycorrhizal fungi on growth, productivity and blast resistance in rice. *Rice* 13, 1–14. doi: 10.1186/s12284-020-00402-7.

Campos-Soriano, L., García-Martínez, J., and San, S.B. (2012). The arbuscular mycorrhizal symbiosis promotes the systemic induction of regulatory defencerelated genes in rice leaves and confers resistance to pathogen infection. *Mol. Plant. Pathol.* 13, 579–592. doi: 10.1111/j.1364-3703.2011.00773.x.

Chan, W.F., Li, H., Wu, F.Y., Wu, S.C., and Wong, M.H. (2013). Arsenic uptake in upland rice inoculated with a combination or single arbuscular mycorrhizal fungi. *J. Hazard Mater.* 15, 1116–1122. doi: 10.1016/j.jhazmat.2012.08.020.

Cornejo, P., Seguel, A., Aguilera, P., Meier, S., Larsen, J., and Borie, F. (2017). Arbuscular mycorrhizal fungi improve tolerance of agricultural plants to cope abiotic stress conditions. In *Plant-Microbe Interactions in Agro-Ecological Perspectives*, eds. D.P. Singh, H.B. Singh, and R. Prabha. Singapore: Springer, pp. 55–80. doi: 10.1007/978-981-10-6593-4_3.

Corrêa, A., Cruz, C., Pérez-Tienda, J., and Ferrol, N. (2014). Shedding light onto nutrient responses of arbuscular mycorrhizal plants: nutrient interactions may lead to unpredicted outcomes of the symbiosis. *Plant Sci.* 221, 29–41. doi: 10.1016/j.plantsci.2014.01.009.

Cosme, M., Stout, M.J., and Wurst, S. (2011). Effect of arbuscular mycorrhizal fungi (*Glomus intraradices*) on the oviposition of rice water weevil (*Lissorhoptrus oryzophilus*). *Mycorrhiza* 21, 651–658. doi: 10.1007/s00572-011-0399-6.

Ding, C., Du, S., Ma, Y., Li, X., Zhang, T., and Wang, X. (2019). Changes in the pH of paddy soils after flooding and drainage: modeling and validation. *Geoderma* 337, 511–513. doi: 10.1016/j.geoderma.2018.10.012.

Fiorilli, V., Vallino, M., Biselli, C., Faccio, A., Bagnaresi, P., and Bonfante, P. (2015). Host and non-host roots in rice: cellular and molecular approaches reveal differential responses to arbuscular mycorrhizal fungi. *Front Plant Sci.* 6, 636. https://doi.org/10.3389/fpls.2015.00636.

Fiorilli, V., Vannini, C., Ortolani, F., Garcia-Seco, D., Chiapello, M., Novero, M., et al. (2018). Omics approaches revealed how arbuscular mycorrhizal symbiosis enhances yield and resistance to leaf pathogen in wheat. *Sci. Rep.* 8, 9625. doi: 10.1038/s41598-018-27622-8.

Gao, X., Kuyper, T.W., Zou, C., Zhang, F., and Hoffland, E. (2007). Mycorrhizal responsiveness of aerobic rice genotypes is negatively correlated with their zinc uptake when non mycorrhizal. *Plant Soil* 290, 283–291.

Gewaily, E.E. (2019). Influence of arbuscular mycorrhizal fungi (AMF) inoculation on the performance of Sakha 107 rice cultivar under different irrigation intervals. *Env. Biodiv. Soil Secur.* 3, 119–129. doi: 10.21608/jenvbs.2019. 15637.1068.

Huang, R., Li, Z., Mao, C., Zhang, H., Sun, Z., Li, H., et al. (2020). Natural variation at OsCERK1 regulates arbuscular mycorrhizal symbiosis in rice. *New Phytol.* 225, 1762–1776. doi: 10.1111/nph.16158.

Konvalinková, T., and Jansa, J. (2016). Lights off for arbuscular mycorrhiza: on its symbiotic functioning under light deprivation. *Front. Plant Sci.* 7, 782. doi: 10.3389/fpls.2016.00782.

Konvalinková, T., Püschel, D., Janoušková, M., Gryndler, M., and Jansa, J. (2015). Duration and intensity of shade differentially affects mycorrhizal growth-and phosphorus uptake responses of Medicago truncatula. *Front. Plant Sci.* 6, 65. doi: 10.3389/fpls.2015.00065.

Maiti, D., Singh, R.K., and Variar, M. (2012). Rice-based crop rotation for enhancing native arbuscular mycorrhizal (AM) activity to improve phosphorus nutrition of upland rice (*Oryza sativa* L.) *Biol. Fertil. Soils* 48(1), 67–73.

Maiti, D., Toppo, N.N., and Variar, M. (2011). Integration of crop rotation and arbuscular mycorrhiza (AM) inoculum application for enhancing AM activity to improve phosphorus nutrition and yield of upland rice (*Oryza sativa* L.). *Mycorrhiza* 21(8), 659–667.

Mitra, D., Uniyal, N., Panneerselvam, P., Senapati, A., and Ganeshamurthy, A.N. (2020). Role of mycorrhiza and its associated bacteria on plant growth promotion and nutrient management in sustainable agriculture. *Int. J Life Sci. Appl. Sci.* 1(1), 1–1.

Mitsuhara, I., Iwai, T., Seo, S., Yanagawa, Y., Kawahigasi, H., Hirose, S., et al. (2008). Characteristic expression of twelve rice PR1 family genes in response to pathogen infection, wounding, and defense-related signal compounds (121/180). *Mol. Genet. Genomics* 279(4), 415–427.

Neuenkamp, L., Zobel, M., Koorem, K., Jairus, T., Davison, J., Pik, M., et al. (2021). Light availability and light demand of plants shape the arbuscular mycorrhizal fungal communities in their roots. *Ecol. Lett.* 24, 426–437. doi: 10.5194/egusphere-egu21-12657.

Panneerselvam, P., Kumar, U., Saha, S., Adak, T., and Munda, S. (2016). Effect of Bispyribac sodium on arbuscular mycorrhizal (AM) fungal association in rice. *NRRI Newsl.* 37(1), 18.

Panneerselvam, P., Kumar, U., Senapati, A., Parameswaran, C., Anandan, A., Kumar, A., Jahan, A., Padhy, S.R., and Nayak, A.K. (2020). Influence of elevated CO_2 on arbuscular mycorrhizal fungal community elucidated using Illumina MiSeq platform in sub-humid tropical paddy soil. *Appl. Soil Ecol.* 145, 103344.

Panneerselvam, P., Kumar, U., Sugitha, T.C.K., Parameswaran, C., Sahoo, S., Binodh, A.K., Jahan, A., and Anandan, A. (2017). Arbuscular mycorrhizal fungi (AMF) for sustainable rice production. In *Advances in Soil Microbiology: Recent Trends and Future Prospects*, eds. T.K Adhya, B. Lal, B. Mohapatra, D. Paul, and S. Das. Singapore: Springer, pp. 99–126.

Panneerselvam, P., Sahoo, S., Senapati, A., Kumar, U., Mitra, D., Parameswaran, C., Anandan, A., Kumar, A., Jahan, A., and Nayak, A.K. (2019). Understanding interaction effect of arbuscular mycorrhizal fungi in rice under elevated carbon dioxide conditions. *J. Basic Microbiol.* 59(12), 1217–1228.

Panneerselvam, P., Saritha, B., Mohandas, S., Upreti, K.K., Poovarasan, S., Sulladmath, V.V., and Venugopalan. R. (2013). Effect of mycorrhiza-associated bacteria on enhancing colonization and sporulation of *Glomus mosseae* and growth promotion in sapota (*Manilkara achras* (mill) Forsberg) seedlings. *Biol. Agric. Hortic.* 29(2), 118–131.

Parniske, M. (2008). Arbuscular mycorrhiza: the mother of plant root endosymbioses. *Nat. Rev. Microbiol.* 6(10), 763–775.

Ray, D.K., West, P.C., Clark, M., Gerber, J.S., Prishchepov, A.V., and Chatterjee, S. (2019). Climate change has likely already affected global food production. *PLoS One* 14, e217148. doi: 10.1371/journal.pone.0217148.

Ruíz-Sánchez, M., Armada, E., Munoz, Y., de Salamone, I.E.G., Aroca, R., Ruız-Lozano, J.M., and Azcon, R. (2011). Azospirillum and arbuscular mycorrhizal colonization enhance rice growth and physiological traits under well-watered and drought conditions. *J. Plant Physiol.* 168(10), 1031–1037.

Sawers, R.J., Ramírez-Flores, M.R., Olalde-Portugal, V., and Paszkowski, U. (2018). The impact of domestication and crop improvement on arbuscular mycorrhizal symbiosis in cereals: insights from genetics and genomics. *New Phytol.* 220, 1135–1140. doi: 10.1111/nph.15152.

Tian, L., Chang, C., Ma, L., Nasir, F., and Tian, C. (2019). Comparative study of the mycorrhizal root transcriptomes of wild and cultivated rice in response to the pathogen magnaporthe oryzae. *Rice* 12, 35. doi: 10.1186/s12284-019-0287-9.

Tie, X.X., Huang, R.J., Dai, W.T., Cao, J.J., Long, X., Su, X.L., et al. (2016). Effect of heavy haze and aerosol pollution on rice and wheat productions in China. *Sci. Rep.* 6, 1–6. doi: 10.1038/srep29612.

Vallino, M., Fiorilli, V., and Bonfante, P. (2014). Rice flooding negatively impacts root branching and arbuscular mycorrhizal colonization, but not fungal viability. *Plant Cell Environ.* 37, 557–572. doi: 10.1111/pce.12177.

Wang, P., Jiang, H., Boeren, S., Dings, H., Kulikova, O., Bisseling, T., et al. (2021). A nuclear-targeted effector of Rhizophagus irregularis interferes with Histone 2B mono-ubiquitination to promote arbuscular mycorrhization. *New Phytol.* 230, 1142–1155. doi: 10.1111/nph.17236.

Wang, Y., Li, T., Li, Y., Björn, L.O., Rosendahl, S., Olsson, P.A., et al. (2015). Community dynamics of arbuscular mycorrhizal fungi in high-input and intensively irrigated rice cultivation systems. *Appl. Environ. Microbiol.* 81, 2958–2965. doi: 10.1128/aem.03769-14.

Wang, Y., Li, Y., Bao, X., Björn, L.O., Li, S., and Olsson, P.A. (2016). Response differences of arbuscular mycorrhizal fungi communities in the roots of an aquatic and a semiaquatic species to various flooding regimes. *Plant Soil* 403, 361–373. doi: 10.1007/s11104-016-2811-7.

Zhang, H., Feng, P., Yang, W., Sui, X., Li, X., and Li, W. (2017). Effects of flooding stress on the photosynthetic apparatus of leaves of two physocarpus cultivars. *J. For. Res.* 29, 1049–1059. doi: 10.1007/s11676-017-0496-2.

Zhang, S., Wang, L., Ma, F., Bloomfield, K.J., Yang, J., and Atkin, O.K. (2015). Is resource allocation and grain yield of rice altered by inoculation with arbuscular mycorrhizal fungi? *J. Plant Ecol.* 8, 436–448. doi: 10.1093/jpe/rtu025.

Zipfel, C., and Oldroyd, G.E. (2017). Plant signalling in symbiosis and immunity. *Nature* 543, 328–336. doi: 10.1038/nature22009.

18 AM Fungi Role in Soil Health Management

Priyanka Adhikari, Kuldeep Joshi, and Pooja Thathola
G.B. Pant National Institute of Himalayan Environment

CONTENTS

18.1 INTRODUCTION

The utmost momentous intimidation soil purpose worldwide are soil destruction, soil organic carbon, unwarranted usage involvement, nutrient unevenness, and unmanageable land supervision activities, i.e., overgrazing, bush sweltering, unremitting crop farming, and tillage cultivation (Yousaf et al., 2016; Wilson et al., 2009). Although the usage of arbuscular mycorrhizal fungi (AMF) as a bio-inoculation has been recognized as an eco-friendly method to expand soil nutrient management, which directly or indirectly increases the soil fertility (Panneerselvam et al., 2012, 2013; Poovarasan et al., 2013; Mohandas et al., 2013; Akhzari et al., 2015). AMF live as obligate symbionts with plants, which necessitate a host, i.e., plant, herbs, and crops, to comprehend their life cycle. AMF play an important role in the enhancement of crop production by enhancing liquid and nutrient uptake, e.g., nitrogen (N), phosphorus (P), and potassium (K) (Anderson et al., 2018). These all-noteworthy characteristics of AMF are because of their mycelium. These mycelia or hyphae engross nutrients through the osmotrophic process and reconnoiter extra superficial zone compared to the

DOI: 10.1201/9781003354086-18

non-mycorrhizal roots. They also produce glomalin and organic acids, which directly help in maintaining the soil structure (Hashem et al., 2018). In fact, AM hyphae can combine soil particles and improve soil structures, increasing soil moisture, soil macropores, water-holding capacity, and porosity (Akhzari et al., 2015). Arbuscular mycorrhizas mycelia improve plant nutrients' acquisition or uptake by increasing the absorbing surface area. Mycorrhizal roots and their exudates are gifted to get supplementary nutrients from nutrient-deficient fields than the non-mycorrhizal roots, the reason to collect a larger capacity of soil than alone roots (Hashem et al., 2018).

Additionally, AMF also payback carbohydrates beginning from the hosts. Previously, lots of workers established that 20% of photosynthetic carbohydrates were attended by AMF in the host plant; furthermore, fats and lipids are the foremost causes of organic carbon, which is transported to the AMF (Luginbuehl et al., 2017). The effects of AMF on soil fertility are described in Figure 18.1. Based on the above discussion, this chapter aims to recapitulate information about the AMF relationship, in specific, the valuable effects on the fertility of the soil. This chapter gives a petite converging on the application of AMF in enlightening soil fertility. The major focus of this chapter is to explore the functions and application of AMF in refining the biochemical, physical, and organic belongings of the soil. Moreover, highlights of few valuable properties of AMF on soil carbon sequestration, soil texture, nutrient stuffing, and the contribution of AMF to the phytochemical composition of adulterated soil are also described.

18.2 IMPACT OF AMF ON SOIL PHYSICAL PROPERTIES

18.2.1 Soil Structure

AMF have an advantageous outcome on the development of soil structure genesis directly and indirectly by numerous biochemical and biophysical process. Their

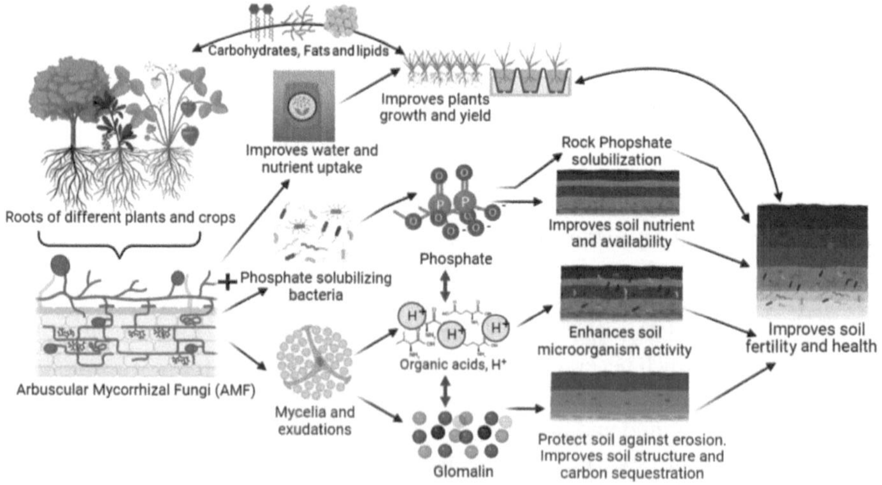

FIGURE 18.1 Paraphernalia of AMF on refining the soil fertility.

mycelia play an imperative role in the soil genesis to a huge extent in the soils (De Novais et al., 2019). At the same time, they are also producing glomalin (glycoprotein), which acts as a soil-binding mediator. Glomalin is aquaphobic and can tolerate the burning temperature of the soil. The aquaphobic nature of the glomalin deliberates and confronts the soil accretions to water uptake (Hashem et al., 2018). We can also say that glomalin performance as an adhesive agent that fixes soil and micronutrients together, which forms a huge and unchanging macro-aggregation, and this aggregation helps against soil erosion and in carbon sequestration (Lehmann et al., 2020).

18.3 ROLE OF AMF ON SOIL CHEMICAL PROPERTIES

18.3.1 Soil Phosphorous

Phosphorous (P) is a significant, non-replaceable component for progress in the agriculture and horticulture field. The build-up of P in soils from manures and composites is known as heritage P (Sattari et al., 2012). The low accessibility of P is the foremost restraint for the lower input in the field of agriculture. AMF play an important role in refining P availability in the soil. Undoubtedly, it is a P activator that can quicken the procedure to transmute P into available forms via a series of chemical responses and biological connections (Zhu et al., 2018). In the case of P deficiency, plants frequently get the advantage from AMF plant symbiosis, and AMF rapidly promote plant development. For sustainable agriculture, the addition of AMF is an essential management practice when the native mycorrhizal potential improves the quantity and quality of soil (Luginbuehl et al., 2017).

18.3.2 Soil Nitrogen

Like P, N is also an imperative element for floras. It is an integral part of coenzymes, phospholipids, and amino acids (Hoeksema et al., 2010). In soil, nitrogen is present in numerous organic forms, i.e., nitrites, nitrates, and ammonium ions. Plant studies validate that extraradical mycelia of AM fungi are clever to obtain inorganic nitrogen either as nitrates or as ammonia, consequently change it to arginine, and convey it to intraradical fungal edifices wherever the amino acid is wrecked down and assimilated into plant proteins (Govindarajulu et al., 2005). AMF were reported for reducing nitrate leaching by 40 times compared to the non-mycorrhizal tomato mutant plants. Moreover, the huge biomass and high N requirement of AMF mean that they symbolize a global N pool to fine roots and play a substantial and overlooked part in the N cycle (Hodge and Fitter, 2010). The roles of AMF in different chemical properties are given in Figure 18.2.

18.3.3 Soil Carbon Cycle and Sequestration

Soil contains higher amounts of C than the collective form of atmosphere and vegetation. Extrametrical mycorrhizal hyphae (EMH) of AMF play a key part

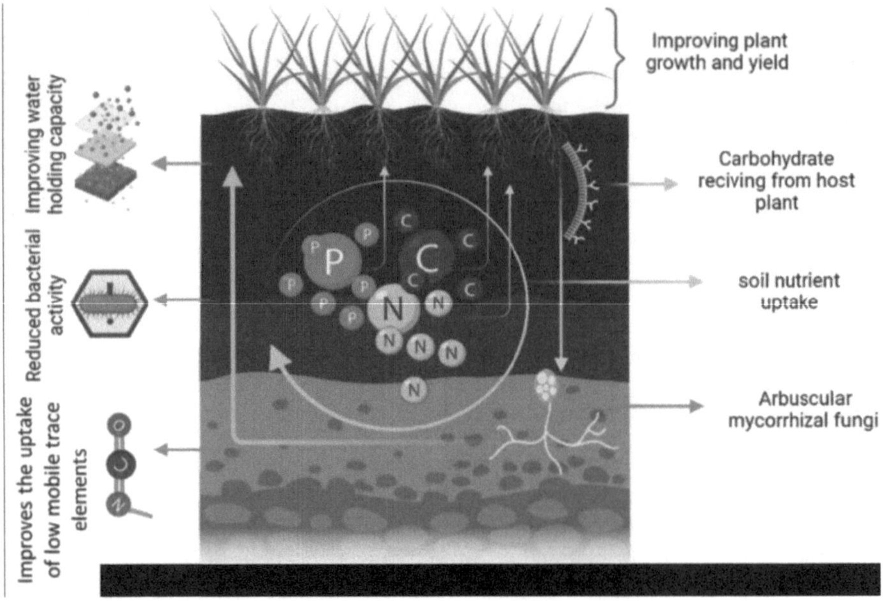

FIGURE 18.2 Roles of AMF on soil chemical properties.

in C translocation hooked on the soil and activate the terrestrial C cycle (Finlay, 2008). Mycorrhizae improve the C sequestration by translocating C after the higher breathing action from place to place near the root and hooking on the soil (Zhu and Miller, 2003). AMF unswervingly effect the soil C sequestration by the turnover, growth, and decomposition of extraradical hyphae in the rhizosphere and soil. AMF play an important role in the global C cycling and consume around 20% of net plant photosynthate. The overall involvement of AMF in soil C sequestration might be contingent on the amount and excellence of hyphae produced, and the oldness and pliability of hyphal deposits (Solaiman, 2014).

18.4 INVOLVEMENT OF AMF IN SOIL TRACE ELEMENT TRANSMISSION

Trace elements play a crucial part in enzymatic happenings related to food-making process in different groups of the plant kingdom (Dominguez-Nuñez et al., 2016). It is well recognized that in soils, AMF permit improved preoccupation of low-slung moveable trace elements, i.e., potassium (K), calcium, magnesium, copper, zinc, iron, manganese, and cobalt (Hashem et al., 2018). Previously, many workers reported on the importance of AMF on trace elements transmission, and the amount of zinc, iron, and manganese is two-fold higher in peanut plants treated with AMF compared to control plants. However, at sometimes when trace elements exist in higher amounts and consequently hold a lethal charisma, in that case, AMF perform a different work by providing defense to the plants and by robust maintenance of these trace elements (Garcia et al., 2016).

18.5 IMPACT OF CROP MANAGEMENT ON AMF

The impact of AMF on crop management should be described as the microorganisms beneficial for the crop's nutrient status, and it should also improve the crop quality. AMF colonized in the plants may exhibit an increase in the level of the secondary metabolites that may improve the antioxidant properties of the plants (Castellanos-Morales et al., 2010). AMF should also increase the dietary crop quality by increasing carotenoid production and certain volatile compounds (Hart et al., 2015). AMF possess a substantial potential for enhancing the yield of the crop. AMF should upsurge the biosynthesis of treasured phytochemicals in plants and assist the plant for vigorous production in the food chain (Sbrana et al., 2014).

18.6 IMPRESSION OF SOIL MANAGING EXERCISE ON AMF

Managing exercises such as cultivation, intensive fertilization, and pesticide application can reduce soil fertility and are the main reason for soil biodiversity (Luginbuehl et al., 2017). For the reduction of such negative effects on soil biodiversity, alternate farming practices are proposed, for obtaining high-quality yields in crops; meanwhile, this should also maintain the soil biodiversity for the long term (Akhzari et al., 2015). The composition of the AMF was designed for long-term organic and conventional farming, and it is not influenced by short-term processes. Although there is no adverse effect of drought on the AMF community, some AMF species are identified as the stress indicators for the AMF community. Furthermore, the different long-term and short-term agricultural practices may not affect the soil biodiversity and AMF community (Hashem et al., 2018).

18.7 CONTRIBUTION OF AMF IN THE PHYTOCHEMICAL OF POLLUTED SOIL

The AMF contributions toward the phytochemical of the polluted soil are broadly believed to support the establishment of plants grown in contaminated soil with different types of pollutants. The contaminants in the soil may accumulate in the plants related to crops, fruits, and vegetables, which may cause various types of health hazards to the plants, animals, and environment (Yousaf et al., 2016). Many reports are published on the connotation of AMF with agricultural practices in the polluted soil which may increase the uptake of essential nutrients and help to remove or degrade the pollutants in the soil or plant root surface (Aguilera et al., 2014). The AMF's strong effect on the development of plants under severely stressful environments is due to the ability of these AMF to increase the processes that are mainly morphological and physiological, which may increase the biomass of the plant and help the plant in the contiguous uptake of important nutrients that are supportive in the development of the plant and assist the plant to survive under the stress condition (Miransari, 2017). The working of the AMF in contaminants removal is that the AMF go to the cell wall of the hyphae or cortical cells, which mainly restrict the uptake of those contaminants, resulting in growth improvement, yield improvement, and nutrient uptake in the plant (Garg and Chandel, 2012).

18.8 CONCLUSION AND FUTURE PERSPECTIVE

The foremost central thing that is present in the soil and that helps in maintaining the soil health and fertility is microorganisms. Among them, the significant one is arbuscular mycorrhizal fungi. They play a major role in balancing plant nutrition, mineral uptake, and water captivation, and at the same time, they provide defense against both biotic and abiotic stresses. In spite of the statistics that the AMF significantly play a refining role in soil health management, understanding of the fundamental processes and pathways still needs a major attention.

- In the future, detailed studies are obligatory to describe the phosphorous and nitrogen threshold in the soil underneath that AMF create interdependence and overhead which AMF are non-lively. Studies on AMF-suitable soil types and optimum-suitable parameters are needed.
- Furthermore, more advanced molecular-level studies are warranted by using advanced molecular tools to examine the directives of nitrogen and its acceptance from the soil all through mycorrhizal association. Extra field experiments are obligatory in the future for testing the influence of AMF diversity on soil and plant diversity.
- A detailed investigation is needed to check the functions of AMF in the transportation of phytohormones from plant-plant and from plant-soil microbe.
- Consequently, research pointing to the identification of such AMF and their bioprospection is much warranted.
- All this research work should be conducted in mutual ways in the following conditions, i.e., laboratory, open and closed greenhouse, field conditions, by using new and novel methods. All these studies will give new heights and milestones to upgrade our good agriculture practices.

ACKNOWLEDGMENTS

The authors are grateful to Director GBP-NIHE, Almora, Uttarakhand, India, for extending the facilities. National Mission on Himalayan Studies (NMHS), Ministry of Environment, Forest & Climate Change, Government of India, New Delhi, is thanked for financial support.

REFERENCES

Aguilera, P., Pablo, C., Fernando, B., and Fritz, O. 2014. Diversity of arbuscular mycorrhizal fungi associated with *Triticum aestivum* L. plants growing in an andosol with high aluminium level. *Agriculture Ecosysytems and Environment* 186:178–184.

Akhzari, D., Attaeian, B., Arami, A., Mahmoodi, F., and Aslani, F. 2015. Effects of vermicompost and arbuscular mycorrhizal fungi on soil properties and growth of *Medicago polymorpha* L. *Compost Science and Utilization* 23(3):142–153.

Anderson, R., Keshwani, D., Guru, A., Yang, H., Irmak, S., and Subbiah, J. 2018. An integrated modeling framework for crop and biofuel systems using the DSSAT and GREET models. *Environmental Modelling and Software* 108:40–50.

Castellanos-Morales, V., Villegas, J., Wendelin, S., Vierheiling, H., Eder, R., and Cardenas-Navorro, R. 2010. Root colonization by the arbuscular mycorrhizal fungus Glomas intraradices alters the quality of strawberry fruits (*Fragaria ananassa* Duch.) at different nitrogen levels. *Journal of Food Science and Agriculture* 90:1774–1782.

De Novais, C. B., Avio, L., Giovannetti, M., De Faria, S. M., Siqueira, J. O., and Sbrana, C. 2019. Interconnectedness, length and viability of arbuscular mycorrhizal mycelium as affected by selected herbicides and fungicides. *Applied Soil Ecology* 143:144–152.

Dominguez-Nuñez, J. A., Benito, B., Berrocal-Lobo, M., and Albanesi, A. 2016. Mycorrhizal fungi: role in the solubilization of potassium. In *Potassium Solubilizing Microorganisms for Sustainable Agriculture*, eds. V. Meena, B. Maurya, J. Verma, and R. Meena. New Delhi: Springer, pp. 77–98.

Finlay, R. D. 2008. Ecological aspects of mycorrhizal symbiosis: with special emphasis on the functional diversity of interactions involving the extraradical mycelium. *Journal of Experimental Botany* 59:1115–1126.

Garcia, K., Doidy, J., Zimmermann, S. D., Wipf, D., and Courty, P. E. 2016. Take a trip through the plant and fungal transportome of mycorrhiza. *Trends in Plant Science* 21:937–950.

Garg, N., and Chandel, S. 2012. Roles of arbuscular mycorrhizal fungi (AMF) on growth, cadmium uptake, osmolyte, and phytochelatin systhesis in *Cajanus cajan* (L.) Mill sp. under NaCl and Cd stresses. *Journal of Plant Growth and Regulation* 31(3):292–308.

Govindarajulu, M., Philip E. P., Hairu J., Jehad A., David D. D., James W. A., Heike B., Peter, J. L., and Yair, S. H. 2005. Nitrogen transfer in the arbuscular mycorrhizal symbiosis. *Nature* 435 (7043):819–823.

Hart, M., Ehret, D. L., Krumbein, A., Leung, C., Murch, S., Turi, C., and Franken, P. 2015. Inoculation with arbuscular mycorrhizal fungi improves the nutritional value of tomatoes. *Mycorrhiza* 25(5):359–376.

Hashem, A., Abd-Allah, E. F., Alqarawi, A. A., Radhakrishnan, R., and Kumar, A. 2018. Plant defense approach of *Bacillus subtilis* (BERA 71) against *Macrophomina phaseolina* (Tassi) Goid in mung bean. *Journal of Plant Interaction* 12:390401.

Hodge, A., and Fitter, A. H. 2010. Substantial nitrogen acquisition by arbuscular mycorrhizal fungi from organic material has implications for N cycling. *Proceedings of the National Academy of Sciences* 107(31):13754–13759.

Hoeksema, J.D. 2010. Ongoing coevolution in mycorrhizal interactions. *New Phytologist*, 187(2):286–300.

Lehmann, J., Hansel, C. M., Kaiser, C., Kleber, M., Maher, K., Manzoni, S., Nunan, N., Reichstein, M., Schimel, J. P., Torn, M. S., Wieder, W. R., and Kogel-Knabner, I. 2020. Persistence of soil organic carbon caused by functional complexity. *Nature Geoscience* 13(529):529–534.

Luginbuehl, L. H., Menard, G. N., Kurup, S., Van Erp, H., Radhakrishnan, G. V., Breakspear, A., Oldroyd, G. E. D., and Eastmond, P. J. 2017. Fatty acids in arbuscular mycorrhizal fungi are synthesized by the host plant. *Science* 356:1175–1178.

Miransari, M. 2017. Arbuscular mycorrhizal fungi and heavy metal tolerance in plants. In *Arbuscular Mycorrhizas and Stress Tolerance of Plants*, ed. Q. S. Wu. Singapore: Springer Nature, pp. 174–161.

Mohandas, S., Poovarasan, S., Panneerselvam, P., Saritha, B., Upreti, K. K., Kamal, R., and Sita, T. 2013. Guava (*Psidium guajava* L.) rhizosphere Glomus mosseae spores harbor actinomycetes with growth promoting and antifungal attributes. *Scientia Horticulturae* 150:371–376.

Panneerselvam, P., Mohandas, S., Saritha, B., Upreti, K. K., Poovarasan, Monnappa, A., and Sulladmath, V. V. 2012. Glomus mosseae associated bacteria and their influence on stimulation of mycorrhizal colonization, sporulation, and growth promotion in guava (*Psidium guajava* L.) seedlings. *Biological Agriculture & Horticulture* 28(4):267–279.

Panneerselvam, P., Saritha, B., Mohandas, S., Upreti, K. K., Poovarasan, S., Sulladmath, V. V., and Venugopalan, R. 2013. Effect of mycorrhiza-associated bacteria on enhancing colonization and sporulation of Glomus mosseae and growth promotion in sapota (Manilkara achras (Mill) Forsberg) seedlings. *Biological Agriculture & Horticulture* 29(2):118–131.

Poovarasan, S., Mohandas, S., Paneerselvam, P., Saritha, B., and Ajay, K. M. 2013. Mycorrhizae colonizing actinomycetes promote plant growth and control bacterial blight disease of pomegranate (*Punica granatum* L. cv Bhagwa). *Crop Protection* 53:175–181.

Sattari, S. Z., Bouwman, A. F., Giller, K. E., and Van Ittersum, M. K. 2012. Residual soil phosphorus as the missing piece in the global phosphorus crisis puzzle. *Proceedings of National Academic Science United States of America* 109:6348–6353.

Sbrana, C., Avio, L., and Giovannetti, M. 2014. Beneficial mycorrhizal symbionts affecting the production of health-promoting phytochemicals. *Electrophoresis* 35:1535–1546.

Solaiman, Z. M. 2014. Contribution of arbuscular mycorrhizal fungi to soil carbon sequestration. In *Mycorrhizal Fungi: Use in Sustainable Agriculture and Land Restoration*, eds. Z. M. Solaiman, L. K. Abbott, and A. Varma. Berlin, Heidelberg: Springer, pp. 287–296.

Wilson, G. W., Rice, C. W., Rillig, M. C., Springer, A., and Hartnett, D. C. 2009. Soil aggregation and carbon sequestration are tightly correlated with the abundance of arbuscular mycorrhizal fungi: results from long term field experiments. *Ecological Letters* 12:452–461.

Yousaf, B., Liu, G., Wang, R., Imtiaz, M., Zia-ur-Rehman, M., Munir, M. A. M., and Niu, Z. 2016. Bioavailablity evaluation, uptake of heavy metals and potential health risks via dietry exposure in urban-industrial areas. *Environmental Science and Pollution Research* 23:22443–22453.

Zhu, J., Li, M., and Whelan, M. 2018. Phosphorus activators contribute to legacy phosphorus availability in agricultural soils: a review. *Science of the Total Environment* 612:522–537.

Zhu, Y. G., and Miller, R. M. 2003. Carbon cycling by arbuscular mycorrhizal fungi in soil-plant systems. *Trends in Plant Science* 8:407–409.

19 OMICS Sciences for Deciphering Plant–Mycorrhizal Symbiosis

Aishwary Purohit
Graphic Era (Deemed to be University)

Debashish Ghosh
CSIR-Indian Institute of Petroleum

Rajesh Kumar
Babasaheb Bhimrao Ambedkar (A Central) University

Amar Jyoti Das
Graphic Era (Deemed to be University)

CONTENTS

19.1 INTRODUCTION

Fungi are filamentous and dwell on various substrates. They majorly depend on the nutritional strategy of the fungal species. Mycorrhizal entities are classified into two broader categories, namely, ectomycorrhiza, those that colonize intercellular spaces in the epidermal root cells, and endomycorrhiza, those that develop inside the roots and cells (Smith and Read, 2010). Ectendomycorrhiza is the third category showing mixed behavior. Mycorrhiza is a ubiquitous mutualistic

DOI: 10.1201/9781003354086-19

TABLE 19.1
Plant and Their Mycorrhizal Counterparts

Family	Species	Fungal Species
Fabaccaea	*Arachis hypogea*	*Glomus ethanicum*
Poaceae	*Triticum peosieum var. fulignosum*	*G. ehtanicum*
Rutaceae	*Citrus aurantum*	*G. intadices*
Fabaccaea	*Vigna unguiculata*	*G. ethanicum*

microorganism and plays an important role in the plant and soil nutrient cycle. Mycorrhiza fungi play a crucial role in a wide range of functions involved in the carbon and nutrient cycles. Among the plant-fungi interaction, symbiotic and neutral association dominates (Tedersoo et al., 2020). Endomycorrhiza can be further classified into vascular arbuscular mycorrhiza, orchidoid mycorrhiza, monotropoid mycorrhiza, ericoid mycorrhiza, arbutoid mycorrhiza, and arbuscular mycorrhiza (AM), which are the most common endomycorrhiza (associated with more than 80% of plant species) and also the most explored ones. *Arbuscule* is a name derived from the Latin word, meaning a bush or a tree-like appearance (Walker, 2013). Hence, the chapter delves into a concept of mycorrhizal symbiosis through OMICS that includes the exploration of interaction in genomics, transcriptomics, proteomics, and metabolomics levels (Table 19.1).

19.2 MYCORRHIZAL SYMBIOSIS

Fungi were not ancient or changed, but evolved just like other eukaryotic organisms. The ecological and morphological diversity of mycorrhizal fungi is matched by their phylogenetic diversity. From rooted saprophytes to rootless gametophytes evolution by DNA-based analyses has predicted a single phylogenetic origin of mycorrhizal fungi to around 353–463 MYA. About 410 MYA, a discovered unequivocal AM fossil from the early Devonian rhynie chart, is associated with *Agalophton major*'s prostat rhizomes (Brundrett, 2002). AM survived throughout the ages. Climate change creates conditions for a new type of symbiotic fungi, the ectomycorrhizal (ECM) symbiosis. Paleontological, paleographical, and molecular evidence pointed to a steep rise in the ECM being proportional to the Pinacae and Angiosperm in the Early Cretaceous, although the records of fossils are observed of Eocene (c. 50 MYA). Some researchers also cited the possibility of fungi promoting colonization of plants on the land (Redecker et al., 2000; Table 19.2).

19.3 COLONIZATION PROCESS AND PRE-SYMBIOTIC SIGNALING

In pre-symbiotic stages, some diffusion molecules (plant root exudates) mediate corresponding concession between the two symbionts. Further, root epidermis

TABLE 19.2
Genes and Their Role in Mycorrhizal Symbiosis

Genes	Function	Phenotypic Response	References
DWARF14	Strigolactone response	Higher level of AM colonization	Yoshida et al. (2012)
miR5229a/b	Target heme peroxidase	Arbuscule development	Branscheid et al. (2011)
DELLA	Transcriptional regulation	Arbuscule development	Floss et al. (2016)
GINI	Fungal spore germination	N/A	
PCT	Pre-symbiotic root branching	Stimulated pre-symbiotic root branching	Hanlon and Coenen (2011)

adheres to the fungal hyphae, and the site of adhesion is known as hyphopodia (Genre et al., 2005). At the time of apparatus formation, defense and cell wall remodeling genes became active (Siciliano et al., 2007). In response to these, fungal hyphae continue to spread and acquire the relevant spaces for further symbiosis.

19.4 MYCORRHIZAL INTERACTION

Biotrophic mutualism starts the interaction of a series of biomolecules with relevant feedback according to their compatibility. Under the presence of inorganic phosphate, plants produce strigolactones (SL), which are carotenoid-derived molecules with hormonal functions in plants and result in the elongation and branching of the mycorrhizal hyphae. SL exposure is responsible for the general activation of the fungal mitochondria (Besserer et al., 2008; Lanfranco et al., 2018). This interaction increases the chitin oligomers by the mycorrhiza. Plant roots exude various significant molecules of small molecular weight in the rhizosphere. This includes secretions of ions, water enzymes, mucilage, and a muster of carbon including primary and secondary metabolites. The host mycorrhizal interaction depends upon the exudates that plant roots release.

19.5 OMICS STUDY OF SYMBIOSIS

The advent of technical advancement in bioinformatics is enabling the research to be conducted on the molecular level to study the interaction of plant mycorrhiza in order to define the nutrient uptake and regulation pathways. This chapter highlights some of the well-observed portions of the fungal OMIC study (Figure 19.1).

19.5.1 Genomics Study of Mycorrhizal Symbiosis

The AM genes are the largest in the fungi, ranging from 153 MB to 1 GB. *Rhizopus irregularis* was sequenced mycorrhizal fungi due to its smaller genome size, i.e., 153 MB frequency of transposable elements (TE) was found higher (at about 36%).

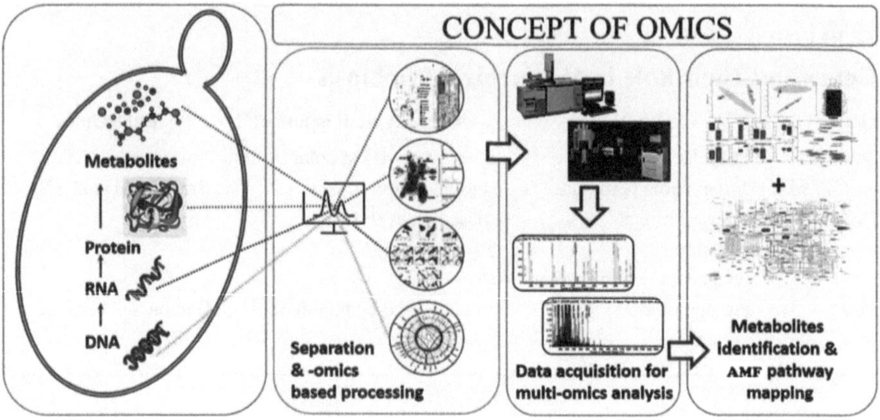

FIGURE 19.1 Concept of OMICS (Sailwal et al., 2020).

Like other mycorrhizal fungi, single-nucleotide polymorphism was in the normal amount. Being the largest fungal genome sequenced, it is suggested to have a relatively smaller proteome when compared to other filamentous fungi (Martin et al., 2010). A study showed that *EcPT4* is the mycorrhiza-specific transport gene, and the level of its expression is proportional to the percentage of root colonization by AM fungi (Pudake et al., 2017). *Mycocosm* is an open-domain fungal genome resource.

19.5.2 Transcriptomics for the Phosphate Uptake

GvPT was the first fungal P*i* (insoluble) transporter from the *Glomus versiforme* to be described (Maldonado-Mendoza et al., 2002). The pursuit of plasma membrane H^+-ATPases is required for the cotransport of H^+ ATPases generates the protein motive driving forces during the uptake of Pi across the extraradial mycelial membrane are encoded by the genes identified in the *R. irregualris* and *G.* mosseae (Ferrol et al., 2000; Requena et al., 2003). A gene-silencing approach was found to be responsible for both development and symbiosis. Irregular arbuscules formation and impaired growth were observed after the GigmPT was inactivated (Maldonado-Mendoza et al., 2001). Various miRNAs involved in mycorrhizal symbiosis are depicted in Table 19.3 (Hanlon and Coenen 2011).

19.5.3 Proteomic Study of the Symbiosis

Effectors are well-known protein molecules of plants that respond to environmental changes. Organic phosphate accounts for the major composition of the rhizosphere. Plants aren't capable of uptaking the organic phosphate directly from the soil. Phosphatase and exophosphatase facilitate the hydrolysis of P*i* from polyphosphates, which are eventually released into the interfacial arbuscular apoplast (Javot et al., 2007). Mycorrhizal factors activate the common symbiotic signaling

TABLE 19.3
miRNAs Involved in Mycorrhizal Symbiosis

Gene	Function	Phenotypic Response	References
miR167	Targets BCP1	Arbuscule development	Pumplin and Harrison (2009)
miR5204	Target GRAS transcription factor	Decrease in mycorrhizal colonization via GRAS	Gobbato et al. (2012)
miR160c	Target auxins response factor gene family	AM symbiosis	Branscheid et al. (2011) and Gobbato et al. (2012)
miR159b	Transcription Efr factors	Targeted silencing led to deformed arbuscules	Branscheid et al. (2011), Devers et al. (2013), and Gobbato et al. (2012)
miR171h	Target NSP2	Increased root colonization and arbuscule formation	Lauressergues et al. (2012)
miR399	Target PHO2 phosphate transporter	Involved in cellular phosphate homeostasis	Branscheid et al. (2010)
miR5282	Targets glutathione-S transferase	Involved in ROS-mediated mycorrhization	

pathway in plants (Parniske, 2008). Several studies have suggested that protein interactions with the pre-symbiotic stage of colonization are not similar to the responses induced in the later stage, i.e., when its association becomes functional. Thus, it can be concluded that the colonization of AM is a non-contemporary process (Simoneau et al., 1994).

19.5.4 Metabolomics

Primary metabolites are affected by the mycorrhizal positively and negatively under environmentally stressed conditions. Strigolactones are essential metabolic compounds ("myc-factors") for the mycorrhizal colonization. Lipochitooligosaccherides are released by AM fungi that act as the inducers (signal) in plants and promote symbiotic-specific response, further assisting the hyphal development (Vierheilig and Piché, 2002), lignans, oxylipins, and certain types of carotenoids. These metabolites were observed to alleviate abiotic stress (salinity, drought, etc.) (Riemann et al., 2015). Steroidal glycoalkaloids are proposed to be antimicrobial (Milner et al., 2011) and insecticidal (Vargas-Ortiz et al., 2018) and are frequently observed in managing the abiotic stress condition.

19.6 CONCLUSION

Overwhelming reports suggest that an AM fungus has a robust symbiotic association with most plant counterparts that benefits both plants and the fungus to a great extent. This chapter highlights the technological advancement that has taken a giant leap in the field of -OMICS in recent times. Although a few reports suggest the success story of -OMICS, still a healthy portion of plant–mycorrhizal interaction, chemical signaling, nutrient regulation pathways, and metabolite synthesis is yet to be discovered. Over the course of time, it will be classified, and eventually, it helps in harnessing the necessities required in the growth and development of plants.

REFERENCES

Besserer, A., Bécard, G., Jauneau, A., Roux, C., & Séjalon-Delmas, N. (2008). GR24, a synthetic analog of strigolactones, stimulates the mitosis and growth of the arbuscular mycorrhizal fungus *Gigaspora rosea* by boosting its energy metabolism. *Plant Physiology*, *148*(1), 402–413.

Branscheid, A., Devers, E. A., May, P., & Krajinski, F. (2011). Distribution pattern of small RNA and degradome reads provides information on miRNA gene structure and regulation. *Plant Signaling & Behavior*, *6*(10), 1609–1611.

Branscheid, A., Sieh, D., Pant, B. D., May, P., Devers, E. A., Elkrog, A., Schauser, L., Scheible, W.-R., & Krajinski, F. (2010). Expression pattern suggests a role of MiR399 in the regulation of the cellular response to local Pi increase during arbuscular mycorrhizal symbiosis. *Molecular Plant-Microbe Interactions: MPMI*, *23*(7), 915–926.

Brundrett, M. C. (2002). Coevolution of roots and mycorrhizas of land plants. *The New Phytologist*, *154*(2), 275–304.

Devers, E. A., Teply, J., Reinert, A., Gaude, N., & Krajinski, F. (2013). An endogenous artificial microRNA system for unraveling the function of root endosymbioses related genes in Medicago truncatula. *BMC Plant Biology*, *13*, 82.

Ferrol, N., Barea, J. M., & Azcón-Aguilar, C. (2000). The plasma membrane H(+)-ATPase gene family in the arbuscular mycorrhizal fungus *Glomus mosseae*. *Current Genetics*, *37*(2), 112–118.

Floss, D. S., Lévesque-Tremblay, V., Park, H.-J., & Harrison, M. J. (2016). DELLA proteins regulate expression of a subset of AM symbiosis-induced genes in Medicago truncatula. *Plant Signaling & Behavior*, *11*(4), e1162369.

Genre, A., Chabaud, M., Timmers, T., Bonfante, P., & Barker, D. G. (2005). Arbuscular mycorrhizal fungi elicit a novel intracellular apparatus in Medicago truncatula root epidermal cells before infection. *The Plant Cell*, *17*(12), 3489–3499.

Gobbato, E., Marsh, J. F., Vernié, T., Wang, E., Maillet, F., Kim, J., Miller, J. B., Sun, J., Bano, S. A., Ratet, P., Mysore, K. S., Dénarié, J., Schultze, M., & Oldroyd, G. E. D. (2012). A GRAS-type transcription factor with a specific function in mycorrhizal signaling. *Current Biology: CB*, *22*(23), 2236–2241.

Hanlon, M. T., & Coenen, C. (2011). Genetic evidence for auxin involvement in arbuscular mycorrhiza initiation. *The New Phytologist*, *189*(3), 701–709.

Javot, H., Pumplin, N., & Harrison, M. J. (2007). Phosphate in the arbuscular mycorrhizal symbiosis: transport properties and regulatory roles. *Plant, Cell & Environment*, *30*(3), 310–322.

Lanfranco, L., Fiorilli, V., & Gutjahr, C. (2018). Partner communication and role of nutrients in the arbuscular mycorrhizal symbiosis. *The New Phytologist*, *220*(4), 1031–1046.

Lauressergues, D., Delaux, P.-M., Formey, D., Lelandais-Brière, C., Fort, S., Cottaz, S., Bécard, G., Niebel, A., Roux, C., & Combier, J.-P. (2012). The microRNA miR171h modulates arbuscular mycorrhizal colonization of Medicago truncatula by targeting NSP2. *The Plant Journal: For Cell and Molecular Biology*, *72*(3), 512–522.

Maldonado-Mendoza, I. E., Dewbre, G. R., & Harrison, M. J. (2001). A phosphate transporter gene from the extra-radical mycelium of an arbuscular mycorrhizal fungus Glomus intraradices is regulated in response to phosphate in the environment. *Molecular Plant-Microbe Interactions: MPMI*, *14*(10), 1140–1148.

Maldonado-Mendoza, I. E., Dewbre, G. R., van Buuren, M. L., Versaw, W. K., & Harrison, M. J. (2002). Methods to estimate the proportion of plant and fungal RNA in an arbuscular mycorrhiza. *Mycorrhiza*, *12*(2), 67–74.

Martin, F., Kohler, A., Murat, C., Balestrini, R., Coutinho, P. M., Jaillon, O., Montanini, B., Morin, E., Noel, B., Percudani, R., Porcel, B., Rubini, A., Amicucci, A., Amselem, J., Anthouard, V., Arcioni, S., Artiguenave, F., Aury, J.-M., Ballario, P., … Wincker, P. (2010). Périgord black truffle genome uncovers evolutionary origins and mechanisms of symbiosis. *Nature*, *464*(7291), 1033–1038.

Milner, S. E., Brunton, N. P., Jones, P. W., O'Brien, N. M., Collins, S. G., & Maguire, A. R. (2011). Bioactivities of glycoalkaloids and their aglycones from Solanum species. *Journal of Agricultural and Food Chemistry*, *59*(8), 3454–3484.

Parniske, M. (2008). Arbuscular mycorrhiza: the mother of plant root endosymbioses. *Nature Reviews Microbiology*, *6*(10), 763–775.

Pudake, R. N., Mehta, C. M., Mohanta, T. K., Sharma, S., Varma, A., & Sharma, A. K. (2017). Expression of four phosphate transporter genes from Finger millet (*Eleusine coracana* L.) in response to mycorrhizal colonization and Pi stress. 3 Biotech, *7*(1), 17.

Pumplin, N., & Harrison, M. J. (2009). Live-cell imaging reveals periarbuscular membrane domains and organelle location in Medicago truncatula roots during arbuscular mycorrhizal symbiosis. *Plant Physiology*, *151*(2), 809–819.

Redecker, D., Kodner, R., & Graham, L. E. (2000). Glomalean fungi from the Ordovician. *Science*, *289*(5486), 1920–1921.

Requena, N., Breuninger, M., Franken, P., & Ocón, A. (2003). Symbiotic status, phosphate, and sucrose regulate the expression of two plasma membrane H+-ATPase genes from the mycorrhizal fungus *Glomus mosseae*. *Plant Physiology*, *132*(3), 1540–1549.

Riemann, M., Dhakarey, R., Hazman, M., Miro, B., Kohli, A., & Nick, P. (2015). Exploring jasmonates in the hormonal network of drought and salinity responses. *Frontiers in Plant Science*, *6*, 1077.

Sailwal, M., Das, A.J., Gazara, R.K., Dasgupta, D., Bhaskar, T., Hazra, S. and Ghosh, D., 2020. Connecting the dots: Advances in modern metabolomics and its application in yeast system. *Biotechnology Advances*, 44, 107616.

Siciliano, V., Genre, A., Balestrini, R., Cappellazzo, G., deWit, P. J. G. M., & Bonfante, P. (2007). Transcriptome analysis of arbuscular mycorrhizal roots during development of the prepenetration apparatus. *Plant Physiology*, *144*(3), 1455–1466.

Simoneau, P., Louisy-Louis, N., Plenchette, C., & Strullu, D. G. (1994). Accumulation of new polypeptides in Ri T-DNA-transformed roots of tomato (*Lycopersicon esculentum*) during the development of vesicular-arbuscular mycorrhizae. *Applied and Environmental Microbiology*, *60*(6), 1810–1813.

Smith, S. E., & Read, D. J. (2010). Mycorrhizal Symbiosis. Academic Press: Cambridge, MA.

Tedersoo, L., Bahram, M., & Zobel, M. (2020). How mycorrhizal associations drive plant population and community biology. *Science*, *367*(6480). doi: 10.1126/science.aba1223.

Vargas-Ortiz, E., Gonda, I., Smeda, J. R., Mutschler, M. A., Giovannoni, J. J., & Jander, G. (2018). Genetic mapping identifies loci that influence tomato resistance against Colorado potato beetles. *Scientific Reports*, *8*(1), 7429.

Vierheilig, H., & Piché, Y. (2002). Signalling in arbuscular mycorrhiza: facts and hypotheses. *Advances in Experimental Medicine and Biology*, *505*, 23–39.

Walker, C. (2013). Arbuscular mycorrhiza in the living collections at the Royal Botanic Garden Edinburgh. *Sibbaldia: The International Journal of Botanic Garden Horticulture*, *11*, 143–157.

Yoshida, S., Kameoka, H., Tempo, M., Akiyama, K., Umehara, M., Yamaguchi, S., Hayashi, H., Kyozuka, J., & Shirasu, K. (2012). The D3 F-box protein is a key component in host strigolactone responses essential for arbuscular mycorrhizal symbiosis. *The New Phytologist*, *196*(4), 1208–1216.

20 AM Fungi
Mass Production, Quality Control, and Application

Sucharita Satapathy and
Shuvendu Shekhar Mohapatra
ICAR-National Rice Research Institute

Puranjoy Sar
Palli Siksha Bhavana (Institute of Agriculture)

Ankita Priyadarshini and Debasis Mitra
ICAR-National Rice Research Institute

Subhadra Pattanayak
Regional Research and Technology Transfer Stations

CONTENTS

DOI: 10.1201/9781003354086-20

20.1 INTRODUCTION: BACKGROUND

The term "mycorrhiza" was taken from a Greek word that implies "fungal root" and was coined by Frank (2005). As it is a mutualistic association between fungal mycelium and plant roots, it provides zinc, phosphorous, and sulfur and helps in transferring and mobilizing various nutrients such as aluminum (Al), copper (Cu), iron (Fe), and potassium (K) from soil to the roots of higher plants (Agnihotri et al., 2021; Akhtar and Abdullah, 2014). They also help in retaining moisture in the root zone and inhibit the growth of soilborne pathogens and nematodes by enhancing the resistance against microbes (Chen et al., 2014, 2018). They possess a vesicle or sac-like structure for storing phosphorous and phospholipids and funneling them into the root system. Also, they exhibit arbuscules, which help in bringing distant nuclei to vesicles and roots. So, this chapter deals with mass production, quality control, and the application of arbuscular mycorrhizal fungi (AMF) for sustainable farming. They are of key importance for sustainable crop improvement.

20.2 MASS PRODUCTION OF AM FUNGI

The AMF enhance soil fertility, soil stability, and soil productivity by improving the soil structure to the degraded soil for which there are many emerging companies like R&D companies that have been working on its mass production (Vasar et al., 2017). The aim is to commercialize, popularize, and use AFM as a biofertilizer. The application of AMF is expanding as AMF play a significant role in the near future mainly in forestry, agriculture, and socioeconomic field (Sharma et al., 2017).

Three well-known systems are widely adopted for the mass production of AMF:

1. Substrate-based production systems
2. Substrate-free production system
3. *In vitro* production system

20.2.1 Substrate-Based Production Systems

This method is well known as the classical method for the mass production of AMF. In this method, first, the plants and their associated symbionts are cultured in sand-based substrate or soil. After the initial creation of AM fungal inoculums, these fungi are propagated for mass multiplication utilizing a single species or a consortium of known AM fungal species in clay or plastic pots or scaled up to medium-size bags and containers and huge raised or grounded beds. The initial inoculums usually consist of a single or a consortium of spores and infected root segments. In order to prepare the after inoculum, the root segments are dehydrated and cut into small and fine pieces to obtain the mixed inoculum, while wet sieving and decanting techniques are used to obtain the single spores. Mixed inoculums are commonly used for the production of those AM fungal species, which may produce intraradical spores and vesicles.

Advantages: This type of system preserves the mass production of single or consortia of AM fungal species, and the nutrient supplies to the AM fungus and plant could be monitored and regulated properly.
Disadvantages: There is the possibility of superfluous contaminants.

20.2.2 Substrate-Free Production Systems

There are various available methodologies, which may vary from each other in aspects of how the nutrient solution is applied and aerated.

20.2.2.1 The Nutrient Flow Technique

It was initially introduced by Mosse and Thompson (1984). This is an alternative system in which the relative area for gas exchange is increased by covering the root zones with a thin nutrient solution.

20.2.2.2 Aeroponics

Aeroponics is a special type of hydroponics system that encourages the soil-free investigations of mycorrhizas by dipping the roots of the host plant and AM fungal propagules in a nutrient medium solution. The spraying of the micro-droplets and the absence of a physical substrate make it more suitable than the hydroponic system.

Advantages: This soil-less system produces substrate-free inoculum.
Disadvantages: Liquid nutrient solutions are highly prone to the growth and development of algal contaminants.

20.2.3 *In vitro* Production Systems

In 1975, Mosse and Hepper first introduced this root organ culture of *in vitro* production system of AMF. This mass-scale production was achieved by root organ culture, where the excised root acts as the host partner. In this method, the freshly isolated mycorrhizal spores were obtained from the rhizospheric soil from trap culture and kept in a beaker. The soil is suspended in tap water. Then, the soil macroaggregate is crushed by hand. Until the upper soil layer becomes transparent, this procedure has to be repeated. After this, AMF spores are obtained and surface sterilized with Chloramine-T for 10 min and then transferred aseptically on Ri-T-DNA. Every 15 weeks, the MS media can be clonally subcultured into two different petri plates, and the formation of extensive clusters of hyphal networks is observed. After months of subcultures, the media plates were incubated at 26°C and transferred to a Falcon tube with 15 Ml citrate buffer. Again they are incubated and the supernatant will be discarded, and spores will are washed, autoclaved, and filtered with a sieve. Then again, they are washed and kept in a –20°C refrigerator for future nursery application. These Ri-T-DNA-transformed hairy roots have given a new impetus to the culture production of these obligate symbionts.

Advantages: The lack of unwanted microorganisms makes this system more appropriate for the mass production of high-quality AM fungal inoculums.

Disadvantages: The risk of cross-contamination increases with an increase in the regular addition of culture media.

20.3 QUALITY CONTROL OF AM FUNGI

Quality control in the formulation of the AMF inoculum is essential for product reliability, uniformity, and reproducibility. The quality control mechanism can be applied in the following way (Salomon et al., 2022).

20.3.1 Laboratory Quality Control

Laboratory quality control is put in action with respect to the production of bacterial spores. They are brought from monospecific spore cultures in the preparation room. The drawn-out spores are then transferred to the petri dishes in the laboratory and emplaced in a refrigerator before going for examination under the stereoscopic microscopes. After that, the examined spores from each petri dish are marked and records are prepared.

20.3.2 Preparation Room Quality Control

This room needs to be detached from the greenhouse and preparation room, whereas the unsterilized soil and media samples must not be reserved in it. Materials like cultures and sterilized growth media are clearly tagged and placed in specified containers. The floor should always be washed improperly, and all the surfaces should be rinsed thoroughly and disinfected. The containers in which the experiments were carried out were surface sterilized with 10% sodium hypochlorite.

20.3.3 Growth Room Quality Control

The temperature of the growth room needs to be controlled, i.e., 22°C. Air is exhausted to the outside and no reprocessing is applied. Surfaces should be colored with antimicrobial paint and decontaminated recurrently, e.g., monthly. All samples were scrutinized for contaminants and pathogens. To keep the samples away from cross-contamination, watering should be done manually.

20.3.4 Storage Room Quality Control

All the preserved samples are placed in plastic bags, with proper labeling or tagging, and the surface of the bags should be rinsed before utilization. Floors and surfaces are cleaned regularly, restraining the generation of dust.

20.3.5 Inoculum Composition and Viability

The selection of AMF species should be relevant to the desired application scenario. If the AMF inoculant is planned for general use, it should contain at least one generalist species that can colonize a broad range of host plant species. The inoculant should contain enough viable propagules to attain AM root colonization. High concentrations of viable propagules are particularly important to account for the declining germination rate of AMF propagules after longer product storage periods. AMF inoculants should be free of plant pathogens and other harmful contaminations. Most of the inoculants are produced *in vivo* on host plants such as sorghum or maize. Here, host plants are grown in sterilized substrates and inoculated with the desired AMF species. These production systems naturally include a range of microorganisms associated with the AMF propagules. However, none of those microorganisms should be pathogenic to the host plant or to its environment.

20.3.6 Carrier Materials

Dispersal of AMF propagules in a carrier material should enable the application of the inoculants without negatively exerting influence on their viability. Various carrier material technologies are accessible for agricultural applications or environmental restoration, such as algal or polymeric beads, liquid solutions, biochar (Sashidhar et al., 2020), and seed coatings. Propagules can be dispersed in coarse material, such as calcinated clay, to facilitate handling. The material should be homogenous so that AMF propagules can be scattered evenly.

20.3.7 Package and Labeling

AMF inoculants should be accumulated in water- and light-proof containers for boosting propagule viability. Caution should be exercised during packaging to ensure that the propagules are unmutilated and inoculum viability is sustained. Distribution channels between producers and consumers must avoid unfavorable conditions, which could harm the propagules, such as prolonged storage times or extreme temperatures below 4°C and above 28°C. The package labeling must comprise all required information about the inoculum content (propagule composition and concentration), production method (*in vivo* or *in vitro*), additives, plant-available nutrients, batch number, production and expiration date, instructions on storage and application, and information about quality measures.

20.3.8 Quality Control: Bioassay

Mycorrhizal inoculants should be tested in a standardized bioassay under controlled conditions. The aim of this bioassay was to assess whether inoculants contain viable propagules and colonize selected host plants in sterilized substrates under controlled conditions. This bioassay provides additional information

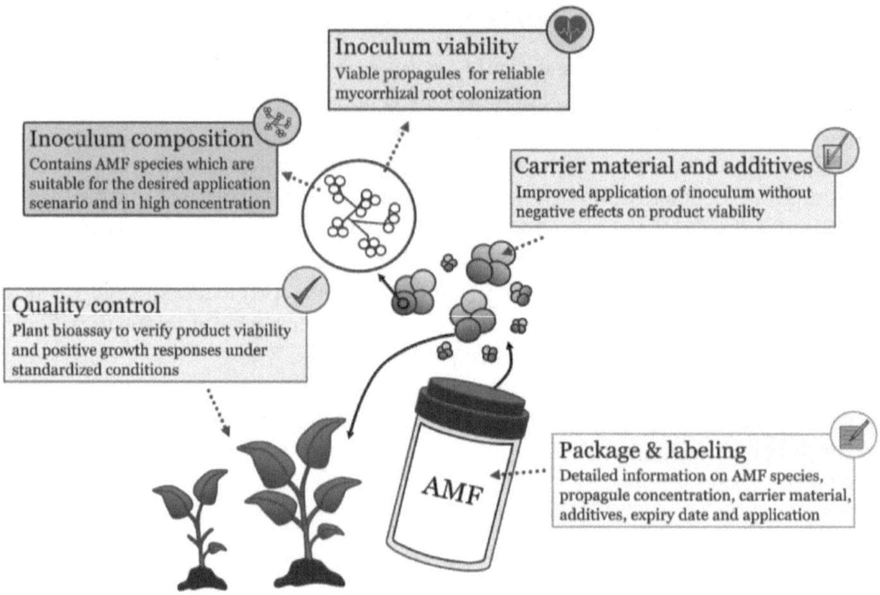

FIGURE 20.1 Quality control and its mechanism.

regarding potential contamination with plant pathogens, which are reflected in the MGR or which can be identified visually (Figure 20.1).

Microbial inoculants are an essential building block for resilient and sustainable food production systems. The growing number of small to medium-size enterprises around the production of AM fungi have seen the market opportunities in coming years. Many researchers have discovered the promises made by these companies about their products, and the results seen by the end users are often world's apart. As natural biological agents and, for AMF at least, non-axenically culturable fungi, there are great problems in presenting the product in the best state for the target markets. Some have taken the approach of single formulations for every market, while others produce a range of products for their target buyers. Whatever the approach, it is increasingly likely that greater regulation and controls over the production and selling of such inocula will be introduced in the coming years.

20.4 APPLICATION OF AM FUNGI

In the natural ecosystem, AMF are used as a biofertilizer because of their wide range of ecological services that improve plant growth and development, soil fertility, stress tolerance, and soil structure. For sustainable agriculture, AMF are receiving attention for their potential use in most crop plants like cereals, vegetables, and fruit trees. The benefit received from AMF depends to a very large extent on the environmental characteristics and the variability of the application of AM Fungi under different environmental conditions indicate that certain combinations are beneficial for the plant, whereas in some cases, the result is neutral or negative (Baar, 2008).

The role of AMF in field crops like wheat, maize, cotton, and soybean is vital. In wheat, the use of AMF as the seed-coating material can reduce the amount of inoculums, increasing efficiency, and cost reduction. The root colonization methodology helps in reducing chemical fertilization consumption, which adds up heavy metals in the soil. Alternatively, the AMF species specificity to the plant genotype is a crucial driver in achieving notable plant response. In the maize production, the shortcomings are nutrient availability, environmental stress like drought, temperature fluctuations, and contaminants. To overcome these limitations, the use of AMF in stress tolerance has been proven as a promising approach. In soybean, mycorrhizal fungi and chemical fertilizer in combination enhances the oil and protein content percentage. The application of mycorrhizal fungi to onions is supposed to be promising for root-disease control, stress amelioration, biocontrol activities, and fruit quality enhancement for human nutrition (Douds et al., 2000). To achieve all these benefits, considering the plant genotype and mycorrhizal fungi combination is important followed by soil properties and inoculation methods. Mycorrhizal inoculation in the plants reduces chemical P_2O_5 requirement in comparison to the non-inoculated plants. In tomatoes, reduced application of P_2O_5 and enhanced mycorrhizal infection may depict the inherited response on tomato growth in low to moderate soil P_2O_5 environment (Sylvia and Chellemi, 2001). The overall application of AMF includes plant defense and disease resistance, protection against abiotic stresses, soil erosion and nutrient leaching, renaturation, reforestation, landscaping, and crop production under water scarcity.

20.5 CONCLUSION

The production of AMF is essential for fundamental research and application purposes. Nowadays, hydroponics, aeroponics, and nutrient flow techniques have been widely used for the mass production of AMF. Among all the currently used techniques, aeroponics is the most appropriate technique for propagating a pure strain of AMF. However, the nutrient flow technique is frequently used for propagating pre-inoculated strains in the host plants, while nutrient film technique or root organ culture techniques have been mostly used for research purposes. Inoculation with AMF may enhance the growth of medicinal plants and the production of medicinal metabolites. However, the diversity, function, and applications of AMF to medicinal plants have received little attention until recently. Advances in studying AMF in association with medicinal plants provide a better understanding of AMF on the improvement of plant growth and active constituent accumulation. The effectiveness of AMF in inducing the medicinal value of plants provides benchmark information that can be used in the development and prioritization of future exploration. This chapter reviews the diversity of AMF in the rhizosphere of some valuable medicinal plants and the impact of AM inoculation on the growth and properties of medicinal plants.

REFERENCES

Agnihotri, R., Bharti, A., Ramesh, A., Prakash, A. and Sharma, M.P., 2021. Glomalin related protein and C16: 1ω5 PLFA associated with AM fungi as potential signatures for assessing the soil C sequestration under contrasting soil management practices. *European Journal of Soil Biology*, 103, p. 103286.

Akhtar, M.S. and Abdullah, S.N.A., 2014. Mass production techniques of arbuscularmycorrhizal fungi: major advantages and disadvantages: a review. *Biosciences Biotechnology Research Asia*, 11, pp. 1199–1204.

Baar, J., 2008. From production to application of arbuscularmycorrhizal fungi in agricultural systems: requirements and needs. In: A. Varma (Ed.), *Mycorrhiza*. Springer: Berlin, Heidelberg, pp. 361–373.

Chen, M., Arato, M., Borghi, L., Nouri, E. and Reinhardt, D., 2018. Beneficial services of arbuscularmycorrhizal fungi–from ecology to application. *Frontiers in Plant Science*, 9, p. 1270.

Chen, Y.L., Li, J.X., Guo, L.P., He, X.H. and Huang, L.Q., 2014. Application of AM fungi to improve the value of medicinal plants. In: Z. M. Solaiman, L. K. Abbott and A. Varma (Eds.), *Mycorrhizal Fungi: Use in Sustainable Agriculture and Land Restoration*. Springer: Berlin/Heidelberg, Germany, pp. 171–187.

Douds, D.D., Pfeffer, P.E. and Shachar-Hill, Y., 2000. Application of in vitro methods to study carbon uptake and transport by AM fungi. *Plant and Soil*, 226(2), pp. 255–261.

Mosse, B. and Thompson, J.P., 1984. Vesicular–arbuscular endomycorrhizal inoculum production. I. Exploratory experiments with beans (Phaseolus vulgaris) in nutrient flow culture. *Canadian Journal of Botany*, 62(7), pp.1523–1530.

Salomon, M.J., Watts-Williams, S.J., McLaughlin, M.J., Bücking, H., Singh, B.K., Hutter, I., Schneider, C., Martin, F., Vosatka, M., Guo, L.D. and Ezawa, T., 2022. Establishing a quality management framework for commercial inoculants containing arbuscularmycorrhizal fungi. *iScience,* 25, p. 104636.

Sharma, S., Sharma, S., Aggarwal, A., Sharma, V., Singh, M.J. and Kaushik, S., 2017. Mass multiplication of arbuscularmycorrhizal fungi. In: A. Aggarwal and K. Yadav (Eds.), *Mycorrhizal Fungi*. Daya Publishing House: New Delhi, pp. 155–174.

Sylvia, D.M. and Chellemi, D.O., 2001. Interactions among root-inhabiting fungi and their implications for biological control of root pathogens. *Advances in Agronomy*, 73, pp. 1–33.

Vasar, M., Andreson, R., Davison, J., Jairus, T., Moora, M., Remm, M., Young, J.P.W., Zobel, M. and Öpik, M., 2017. Increased sequencing depth does not increase captured diversity of arbuscularmycorrhizal fungi. *Mycorrhiza*, 27(8), pp. 761–773.

21 AM Fungi Production Upscaling, Government Regulations, Marketing, and Commercialization

Wiem Alloun
University of Constantine 1

Somya Sinha
Graphic Era (Deemed to be) University

Debasis Mitra
Raiganj University

CONTENTS

21.1 INTRODUCTION

Abiotic stressors affect the yield and development of plants (Mitra et al., 2021). Abiotic factors that affect crop growth and conventional farming methods, such as the excessive use of chemical fertilizers and agro-actives, have a negative impact on the ecosystem. To increase crop output, arbuscular mycorrhizal fungi (AMF), among other environmentally friendly management techniques, are critical (Panneerselvam et al., 2017, 2019).

DOI: 10.1201/9781003354086-21

The term "biofertilizer" is frequently used to describe AMF (Mitra et al., 2020; Pirttilä et al., 2021). In a few studies, AMF were found to improve plant growth under difficult circumstances. Additionally, it is generally recommended that the administration of AMFs to host plants endows them with tolerance to various stressful circumstances, such as heat, salinity, drought, metals, and extremely high temperatures (Panneerselvam et al., 2020). By encouraging tolerance mechanisms and preventing the suppression of essential metabolic processes, AMF may benefit host plants. By supplying host plants with essential inorganic plant nutrients, AMF serve as natural root symbionts that improve growth and yield in both unrestricted and restricted environments (Sahoo et al., 2017).

AMF's potential as a bio-inoculant could improve plants' capacity to adapt to changing environmental conditions (Mitra et al., 2022). It is necessary to pursue research goals that concentrate on the AMF-mediated enhancement of crop quality and yield. It is essential to promote the use of AMF if current global agricultural systems are to remain consistently sustainable. Without any doubt, using AMF to enhance agriculture can significantly reduce the need for synthetic fertilizers and other chemicals, promoting bio-healthy farming. This chapter contains comprehensive and current information. The production of AMF on an industrial scale, their incorporation into modern agricultural biosystems, their advantages and applications, and governmental control over their commercialization are all covered.

21.2 FROM PLANT SYMBIONTS TO BIO-INOCULANTS

Biofertilizers are made up of naturally occurring ingredients and are used to improve soil fertility (Pohanka and Vlcek, 2020). These fertilizers work wonders to improve soil quality and encourage plant growth. Numerous studies on AMF conducted over the past 20 years have found several benefits for soil health and agricultural productivity. AMF may eventually replace inorganic fertilizers because the mycorrhizal application can successfully reduce the quantitative usage of chemical fertilizer input, particularly phosphorus.

The persistent use of synthetic agro-actives has caused problems for plants, soil, and human health in addition to deteriorating food quality, soil health, and air and water systems (Yang et al., 2004). AMF are expected to cut the need for chemical fertilizers by up to 50% for the best agricultural output. However, the range of plants and typical stressful environments will affect this estimate. AMF can enhance crop plant growth and productivity, potentially assisting in meeting the rising demand for food brought on by the world's expanding population. Environmentally friendly technologies will also be vigorously promoted due to their widespread use.

Identifying genes and gene products regulating AMF-mediated growth and development modulation in response to stressful *stimuli* should be the primary focus of future studies. Prospects in this wide research field should concentrate on identifying host and AMF-specific protein components that regulate symbiotic interactions and the primary cellular and metabolic pathways under various environmental conditions (Posta and Duc, 2020). Crop productivity can be increased by comprehending how AMF-induced modulations in tolerance mechanisms and the crosstalk activated to govern plant performance. To better understand their function in nature as a biofertilizer for sustainable agricultural production, AMF must be investigated at all levels (Begum et al., 2019).

21.3 UPSCALING THE APPLICATION OF AMF IN AGRICULTURE

AMF offer promising opportunities for use in agricultural production under various conditions because of the many benefits they bring to their hosts. Since most crops are AMF hosts, it might be advantageous to inoculate them with the illness. In fact, several studies have shown the advantages of using commercial AMF inoculum on crops under agricultural conditions (Weber, 2014).

Numerous studies have shown that AMF enhance plant health and productivity yield (Hijri, 2016; Rouphael et al., 2015; Bell et al., 2016). AMF enhance plant nutrition by absorbing and transporting mineral elements outside the plant rhizosphere's depletion zones. They also change secondary metabolism, raising the product's nutritional value. AMF also affect the hormonal balance of the host plant, inhibiting growth and enhancing resistance to abiotic stressors (Rouphael et al., 2015). One crucial aspect of this is to induce the development of the root system (Gutjahr and Paszkowski, 2013; Chen et al., 2018).

21.3.1 Agricultural and Horticultural Production

The application of AM is particularly intriguing for high-value crops, such as horticulture, and the adaptation of cuttings and micro-propagated plantlets in nurseries. This may be due to the labor-intensive nature of the preparation and deployment of the AM fungal inoculum. Sterile *in vitro* micropropagation is significant in horticultural plants. Thus, the transfer to the soil of the plantlets may be of substantial loss (Schubert and Lubraco, 2000).

The inoculation of fruit trees with AMF can hasten the transition between micropropagation and transplantation and enhance the health of the plantlets, increasing plant production's efficiency with improved commercial features (Vestberg et al., 2002). Apple and peach cuttings that get more strength from AM fungal inoculum are a great illustration of such an application (Chen et al., 2018).

Regardless of whether the potential of these symbiotic fungi has been established on a laboratory scale, work must be carried out to implement them effectively and widely in agricultural systems (Chenchouni, Mekahlia, and Beddiar, 2020).

21.3.2 Large Agricultural Scale Application of AMF

When determining appropriate combinations and inoculation techniques to apply beneficial bacteria, scaling up from laboratory testing, greenhouse research, and eventually to field conditions is difficult (Lugtenberg and Kamilova, 2009). Seed coating can introduce the AM fungus to several crops, ensures the connection of AM fungal propagules with developing roots, and guarantees colonization at an early stage of plant development.

The positive effects of microbial inoculation obtained under greenhouse experiments are not always transferable to field trials. Thus, it is essential to understand the biases of inoculation performance (Rocha et al., 2019). However, studies that consider more than one experimental scale are still rare. As far as we know, there have not been any field investigations on seed-covering inoculation with AMF.

Large-scale agricultural plant production can benefit from arbuscular mycorrhizal inoculation as well. One particularly well-documented case is a substantial increase in tuber production following inoculation with the commercial strain *R. irregularis* (DAOM, 197198) in a significant meta-analysis of potato yield in 231 field experiments in Europe and North America (Hijri, 2016). Interestingly, in each field test, the farmers handled the application and evaluation following their conventional agricultural practice (including applying pesticides and fertilizers).

The replicate plots were excluded from the experimental design, and the experimental settings were heterogeneous because of this method. However, the numerous field locations gave highly encouraging results and robustness. Interestingly, a general favorable effect was seen regardless of the location, type of soil, experimenter, and specific agricultural techniques. The average production improvement in these 231 field trials was 3.9 tons/ha, or 9.5% of the overall crop output. Nearly 80% of all trials were lucrative due to the use of AMF, with an estimated threshold for the profitability of 0.67 tons/ha improved yield (Chen et al., 2018).

This excellent meta-analysis suggests that AM can significantly increase the income of potatoes and possibly other crop growers. AMF use can cut the need for fertilization without affecting yields, increasing profitability even further. Such large-scale experiments produce more reliable results than more controlled greenhouse or small-scale trials (Rocha et al., 2019; Chen et al., 2018).

Although there is much promise for AM applications in horticulture and agriculture, many outside factors need to be considered for AMF in extended field applications to be valuable and successful. Extensive fertilizer application, particularly P, interferes with the colonization and abundance of AMF (Rocha et al., 2019; Hartmann et al., 2015).

Additionally, using particular biocides and cropping with non-host plants impact AMF symbiosis (Njeru et al., 2015). Additionally, AM fungus selection should be based on the crop specie of the host plant to which they are symbiotic because they might not offer the same benefits to another host (Chen et al., 2018).

21.4 COMMERCIAL USE OF AM FUNGI

Because of AMF benefits, the markets for AM-related properties have significantly grown over the past few decades, expanding the pool of performers, the range of products, and the market size. Privately owned, generally small companies constitute most of the AM-related industry. Thus, only little information about the dynamics of market worldwide shares is available to the public. The major actors are based in North America, Europe, Asia, and Latin America. The region's top markets are the United States, Canada, Mexico, Brazil, Argentina, Colombia, and Chile. India, followed by China, is the largest power in Asia (Chen et al., 2018). The European market, including Belgium, Estonia, and Switzerland, is one of the largest for mycorrhizal biostimulants in 2017, with more than 75 businesses commercializing AM-based products for horticulture, agriculture, forestry, and soil remediation (Chen et al., 2018). On an agricultural scale, mycorrhizal inoculation is far less expensive, with an estimated investment of 135 $ per hectare in the case of potatoes in the United States (Hijri, 2016). Many products also contain mixed fungal inocula, sometimes in conjunction with ectomycorrhizal fungi or plant growth-promoting rhizobacteria.

Over the past decade, the Indian market has shown exceptional growth because the Indian government promoted mycorrhiza-based biostimulants, with the efforts undertaken by The Energy and Resources Institute (TERI).

21.5 GOVERNMENT REGULATION AND QUALITY CONTROL OF COMMERCIAL AMF-BASED PRODUCTS

It has been reported that the native AMF species may fail to provide the desired benefits under local conditions (Bender et al., 2019), necessitating intervention with commercial AMF inputs. The leading cause may be the harsh agricultural practices that disrupt indigenous soil microflora. Specific inoculant characteristics must be adequately regulated and controlled to prepare commercial AMF inoculants in the field and make decisions on potential investment in the commercial area. These features include the quantity, quantity, and viability of propagules; the infectivity of strains; the composition of the inoculant, whether single or consortia of AMF species; or combined with other microbes (Salomon et al., 2022). Cost-effective testing must be performed on inoculants to prevent potential contaminations and assess their efficiency. Manufacturers should also prescribe situations that call for re-inoculation after the initial application and deliver an economically and environmentally effective dosage.

Molecular tools such as quantitative real-time PCR and next-generation sequencing (Jack et al., 2021), and other methods continue to be indispensable for assessing the invasiveness, persistence, and effectiveness of large-scale inoculum production in field trials (Stefani et al., 2020). Nevertheless, regulatory institutions must be reinforced, particularly in developing nations (in Asia and Africa).

21.6 CONCLUSION

To select or develop effective combinations and application strategies, the process must be scaled up from laboratory investigations to greenhouse studies to field conditions. This requires knowledge of the target site's mycorrhizal potential. Studies that consider various experimental scales are uncommon. However, microbial inoculation's beneficial effects aren't always replicated in the field under field conditions. Understanding the biases involved in the immunization performance is essential for this purpose. Its effectiveness is influenced by the quality of the bioproducts itself, agricultural practices, and environmental factors. Thus, additional regulation and control investigations are needed to evaluate the effectiveness of these bio-inoculants, their invasive nature, and their evolution in real time at a particular application location.

REFERENCES

Begum, N., C. Qin, M. A. Ahanger, S. Raza, M. Ishfaq Khan, M. Ashraf, N. Ahmed, and L. Zhang. 2019. "Role of Arbuscular Mycorrhizal Fungi in Plant Growth Regulation: Implications in Abiotic Stress Tolerance." *Frontiers in Plant Science* 10. https://doi.org/10.3389/FPLS.2019.01068.

Bell, T. H., F. O. P. Stefani, K. Abram, J. Champagne, E. Yergeau, M. Hijri, and M. St-Arnaud. 2016. "A Diverse Soil Microbiome Degrades More Crude Oil than Specialized Bacterial Assemblages Obtained in Culture." *Applied and Environmental Microbiology* 82 (18): 5530–41. https://doi.org/10.1128/AEM.01327-16.

Bender, S. F., K. Schlaeppi, A. Held, and M. G. A. Van der Heijden. 2019. "Establishment Success and Crop Growth Effects of an Arbuscular Mycorrhizal Fungus Inoculated into Swiss Corn Fields." *Agriculture, Ecosystems and Environment* 273: 13–24. https://doi.org/10.1016/j.agee.2018.12.003.

Chen, M., M. Arato, L. Borghi, E. Nouri, and D. Reinhardt. 2018. "Beneficial Services of Arbuscular Mycorrhizal Fungi – from Ecology to Application." *Frontiers in Plant Science*, Frontiers Media S.A. https://doi.org/10.3389/fpls.2018.01270.

Chenchouni, H., M. N. Mekahlia, and A. Beddiar. 2020. "Effect of Inoculation with Native and Commercial Arbuscular Mycorrhizal Fungi on Growth and Mycorrhizal Colonization of Olive (*Olea Europaea* L.)." *Scientia Horticulturae* 261: 108969. https://doi.org/10.1016/j.scienta.2019.108969.

Gutjahr, C., and U. Paszkowski. 2013. "Multiple Control Levels of Root System Remodeling in Arbuscular Mycorrhizal Symbiosis." *Frontiers in Plant Science*, Frontiers Research Foundation. https://doi.org/10.3389/fpls.2013.00204.

Hartmann, M., B. Frey, J. Mayer, P. Mäder, and F. Widmer. 2015. "Distinct Soil Microbial Diversity under Long-Term Organic and Conventional Farming." *ISME Journal* 9 (5): 1177–94. https://doi.org/10.1038/ismej.2014.210.

Hijri, M. 2016. "Analysis of a Large Dataset of Mycorrhiza Inoculation Field Trials on Potato Shows Highly Significant Increases in Yield." *Mycorrhiza* 26 (3): 209–14. https://doi.org/10.1007/s00572-015-0661-4.

Jack, C. N., R. H. Petipas, T. E. Cheeke, J. L. Rowland, and M. L. Friesen. 2021. "Microbial Inoculants: Silver Bullet or Microbial Jurassic Park?" *Trends in Microbiology*. https://doi.org/10.1016/j.tim.2020.11.006.

Lugtenberg, B., and F. Kamilova. 2009. "Plant-Growth-Promoting Rhizobacteria." *Annual Review of Microbiology*. https://doi.org/10.1146/annurev.micro.62.081307.162918.

Mitra, D., B. E. Guerra Sierra, B. Khoshru, S. De Los Santos Villalobos, C. Belz, P. Chaudhary, F. N. Shahri, et al. 2021. "Impacts of Arbuscular Mycorrhizal Fungi on Rice Growth, Development, and Stress Management with a Particular Emphasis on Strigolactone Effects on Root Development." *Communications in Soil Science and Plant Analysis* 52 (14): 1591–1621. https://doi.org/10.1080/00103624.2021.1892728.

Mitra, D., B. Saritha, E. Janeeshma, P. Gusain, B. Khoshru, F. A. Abo Nouh, A. Rani, et al. 2022. "Arbuscular Mycorrhizal Fungal Association Boosted the Arsenic

Resistance in Crops with Special Responsiveness to Rice Plant." *Environmental and Experimental Botany*. https://doi.org/10.1016/j.envexpbot.2021.104681.

Mitra, D., N. Uniyal, P. Panneerselvam, A. Senapati, A. N. Ganeshamurthy, and D. Jain. 2020. "Role of Mycorrhiza and Its Associated Bacteria on Plant Growth Promotion and Nutrient Management in Sustainable Agriculture." *International Journal of Life Sciences and Applied Sciences* 1 (1): 1–1. https://www.ijlsas.com/?mno=302644881.

Njeru, E. M., L. Avio, G. Bocci, C. Sbrana, A. Turrini, P. Bàrberi, M. Giovannetti, and F. Oehl. 2015. "Contrasting Effects of Cover Crops on 'Hot Spot' Arbuscular Mycorrhizal Fungal Communities in Organic Tomato." *Biology and Fertility of Soils* 51 (2): 151–66. https://doi.org/10.1007/s00374-014-0958-z.

Panneerselvam, P., S. Sahoo, A. Senapati, U. Kumar, D. Mitra, C. Parameswaran, A. Anandan, A. Kumar, A. Jahan, and A. K. Nayak. 2019. "Understanding Interaction Effect of Arbuscular Mycorrhizal Fungi in Rice under Elevated Carbon Dioxide Conditions." *Journal of Basic Microbiology* 59(12), 1217–28.

Panneerselvam, P., U. Kumar, A. Senapati, C. Parameswaran, A. Anandan, A. Kumar, A. Jahan, S. R. Padhy, and A. K. Nayak. 2020. "Influence of Elevated CO_2 on Arbuscular Mycorrhizal Fungal Community Elucidated Using Illumina MiSeq Platform in Sub-Humid Tropical Paddy Soil." *Applied Soil Ecology* 145, 103344.

Panneerselvam, P., U. Kumar, T.C.K. Sugitha, C. Parameswaran, S. Sahoo, A. K., Binodh, A. Jahan, and A. Anandan. 2017. Arbuscular mycorrhizal fungi (AMF) for sustainable rice production. In: T. K. Adhya, B. Lal, B. Mohapatra, D. Paul, and S. Das (Eds.), *Advances in Soil Microbiology: Recent Trends and Future Prospects* (pp. 99–126). Springer, Singapore.

Pirttilä, A. M., H. M. P. Tabas, N. Baruah, and J. J. Koskimäki. 2021. "Biofertilizers and Biocontrol Agents for Agriculture: How to Identify and Develop New Potent Microbial Strains and Traits." *Microorganisms*, MDPI. https://doi.org/10.3390/microorganisms9040817.

Pohanka, M., and V. Vlcek. 2020. "Immunoassay of Glomalin by Quartz Crystal Microbalance Biosensor Containing Iron Oxide Nanoparticles." *International Journal of Analytical Chemistry* 2020. https://doi.org/10.1155/2020/8844151.

Posta, K., and N. H. Duc. 2020. "Benefits of Arbuscular Mycorrhizal Fungi Application to Crop Production under Water Scarcity." *Drought Detection and Solution*, 1, 1–26.

Rocha, I., I. Duarte, Y. Ma, P. Souza-Alonso, A. Látr, M. Vosátka, H. Freitas, and R. S. Oliveira. 2019. "Seed Coating with Arbuscular Mycorrhizal Fungi for Improved Field Production of Chickpea." *Agronomy* 9 (8): 471. https://doi.org/10.3390/agronomy9080471.

Rouphael, Y., P. Franken, C. Schneider, D. Schwarz, M. Giovannetti, M. Agnolucci, S. De Pascale, P. Bonini, and G. Colla. 2015. "Arbuscular Mycorrhizal Fungi Act as Biostimulants in Horticultural Crops." *Scientia Horticulturae*, Elsevier. https://doi.org/10.1016/j.scienta.2015.09.002.

Sahoo, S., P. Panneerselvam, T. Chowdhury, A. Kumar, U. Kumar, A. Jahan, A. Senapati, and A. Anandan. 2017. "Understanding the AM Fungal Association in Flooded Rice under Elevated CO_2 Condition." *ORYZA-An International Journal on Rice* 54(3), 290–7.

Salomon, M. J., R. Demarmels, S. J. Watts-Williams, M. J. McLaughlin, A. Kafle, C. Ketelsen, A. Soupir, H. Bücking, T. R. Cavagnaro, and M. G.A. van der Heijden. 2022. "Global Evaluation of Commercial Arbuscular Mycorrhizal Inoculants under Greenhouse and Field Conditions." *Applied Soil Ecology* 169. https://doi.org/10.1016/j.apsoil.2021.104225.

Schubert, A., and G. Lubraco. 2000. "Mycorrhizal Inoculation Enhances Growth and Nutrient Uptake of Micropropagated Apple Rootstocks during Weaning in

Commercial Substrates of High Nutrient Availability." *Applied Soil Ecology* 15(2): 113–18. https://doi.org/10.1016/S0929-1393(00)00086-X.
Stefani, F., S. Dupont, M. Laterrière, R. Knox, Y. Ruan, C. Hamel, and M. Hijri. 2020. "Similar Arbuscular Mycorrhizal Fungal Communities in 31 Durum Wheat Cultivars (*Triticum Turgidum* L. Var. Durum) under Field Conditions in Eastern Canada." *Frontiers in Plant Science* 11. https://doi.org/10.3389/fpls.2020.01206.
Vestberg, M, A. C. Cassells, A. Schubert, C. Cordier, and S. Gianinazzi. 2002. "Arbuscular Mycorrhizal Fungi and Micropropagation of High Value Crops." In: S. Gianinazzi, H. Schüepp, J. M. Barea, and K. Haselwandter (Eds.), *Mycorrhizal Technology in Agriculture* (pp. 223–33). Birkhäuser, Basel. https://doi.org/10.1007/978-3-0348-8117-3_18.
Weber, O. B. 2014. "Biofertilizers with Arbuscular Mycorrhizal Fungi in Agriculture." In: Z. M. Solaiman, L. K. Abbott, and A. Varma (Eds.), *Mycorrhizal Fungi: Use in Sustainable Agriculture and Land Restoration* (pp. 45–66). Springer. https://doi.org/10.1007/978-3-662-45370-4_4.
Yang, S., F. Li, S. S. Malhi, P. Wang, D. Suo, and J. Wang. 2004. "Long-Term Fertilization Effects on Crop Yield and Nitrate Nitrogen Accumulation in Soil in Northwestern China." *Agronomy Journal* 96(4), 1039–49. https://doi.org/10.2134/agronj2004.1039.

22 Rice Seed Priming with AMF and AMF-Associated Bacteria for Crop Enhancement

R. Djebaili and B. Farda
University of L'Aquila

G. Capoani
University of Milan

G. Pagnani
University of Teramo

M. Pellegrini
University of L'Aquila

CONTENTS

22.1 INTRODUCTION: BACKGROUND AND RATIONALE

Asian agriculture mostly depends on rice (*Oryza sativa* L.) and rice-based systems. With a yearly production of about 500 million tons, it is the second most produced crop in the world after maize (Shasmita et al., 2022). Nevertheless, rice productivity is decreasing for the wide use of agrochemicals in agriculture and climate change. These factors significantly contribute to plant development's abiotic and biotic stresses (Majeed, Muhammad, and Ahmad, 2018).

DOI: 10.1201/9781003354086-22

Modern agriculture practices and technology must be enhanced to raise crop yield and improve agricultural output under stressful conditions. The main challenges are increasing food consumption, maintaining worldwide competitiveness, and providing high-quality agricultural products.

Among innovative and sustainable technologies, seed priming methods seem to be valuable approaches. The practice of "priming" is well known for improving the quality of seeds. It is a pre-sowing method that involves moistening seeds enough to trigger pre-germination metabolic pathways. It induces seeds to germinate more quickly, thus increasing agricultural yields and biotic/abiotic stress resistance (Saha et al., 2022). "Priming" includes conventional (hydro-priming, osmo-priming, chemo-priming, and biopriming) or advanced techniques (nano-priming and using of physical agents) (do Espirito Santo Pereira et al., 2021).

A biopriming method is a treatment in which seeds are treated with a combination of living microbial inoculum and bioactive compounds to improve their germination, plant growth, and qualitative crop production (Sarkar et al., 2021). It is well recognized that certain microorganisms can form endophytic associations with plants, promoting plant growth, producing phytohormones, and enhancing resilience to biotic and abiotic stress (Arif, Batool, and Schenk, 2020). This chapter investigates the potential application of arbuscular mycorrhizal fungi (AMF) and AMFb as a rice seed biopriming strategy to improve crop productivity and outputs under stressful conditions.

22.2 AMF AND AMF-ASSOCIATED BACTERIA

Among the biopriming microbial inoculants, AMF establishes symbiotic relationships with plants, improving water and mineral nutrition, controlling root exudates flow, acidifying soil pH, and enhancing plant nutrition (Mitra et al., 2021). AMF also induce plant tolerance to abiotic and biotic stresses by changing root architectural structure (Behrooz et al., 2019). It is well known that a third symbiotic partner may influence the capacity and effectiveness of AMF colonization. AMFb, mainly belonging to *Bacillus* genus and several genera of Actinobacteria, are usually found in AMF structures (Ujvári et al., 2021). Various and numerous bacterial communities can be found in the mycorrhizosphere or the region around mycorrhizal roots, spores, and extraradical hyphae (Sangwan and Prasanna, 2022). This microbiota revealed a variety of functional activities, from "mycorrhiza helper" (MH) to "plant growth promoting" (PGP) functions (Wagner et al., 2019). PGP microbiota improve plant growth, nutrition, and health. Simultaneously, MH microbiota enhance spore germination, mycelial development, and the creation of mycorrhiza. Furthermore, certain elements of such a beneficial microbiome exhibit both MH and PGP characteristics (Ujvári et al., 2021). A summary of the AMF and AMFb interactions with plants relevant to agricultural applications is shown in Figure 22.1.

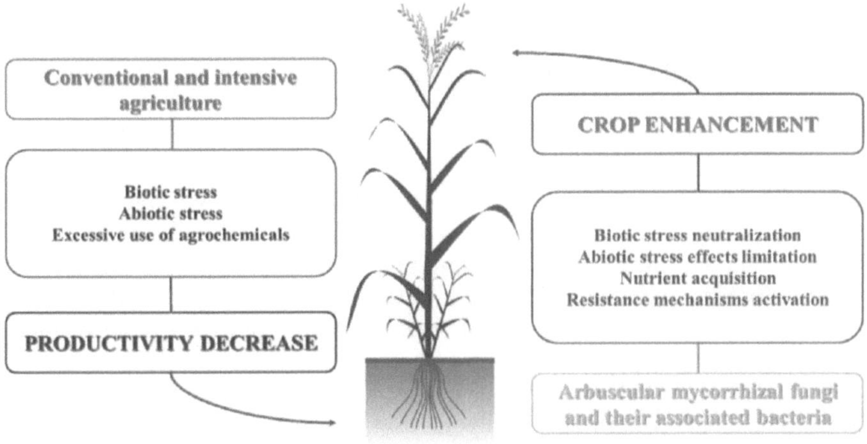

FIGURE 22.1 Summary of the AMF and AMF-associated bacteria interactions with plants useful in plant growth and biotic and abiotic stress tolerance.

22.3 ROLE OF AMF AND AMF-ASSOCIATED BACTERIA IN NUTRIENTS UPTAKE AND PLANT GROWTH

As the focus of crops with agronomical interest has always been on productivity, one of the key features of how a plant can grow with a better yield or with better nutritional values is the nutrient uptake that can lead to optimal plant growth. For this purpose, researchers have been investigating the role and potential application of AMF and AMFb to enhance nutrient uptake efficiency.

As the root apparatus is physically limited to a certain extent, the network that originates between the plant and AMF deviates from the plant spectrum of nutrients. The depletion area is exceeded, and the AMF can reach parts of the soil where the nutrients are still available. For plants and fungi, symbiosis confers advantages explaining why sometimes it is easy to detect in natural conditions the rising of this bond (Bernaola et al., 2018). Nonetheless, the benefit coming from AMF is not restricted to a matter of extension. However, the fungi can translocate and help absorb such elements within the plant. Most of the current studies investigate the mechanisms and the improvement of N and P uptake (Luginbuehl and Oldroyd, 2017). This aspect is in line with the increasing concerns about the availability of P and also about the environmental effects of these elements on the ecosystems (Giovannini et al., 2020).

On the contrary, the requirements for rice cultivation are, without a doubt, increasing the complexity of this relationship. However, most AMF having a role in plant growth improvement are mainly of the genus *Glomus*. *Rhizophagus irregularis* is part of this large group of microorganisms, and it is used as a model to unveil the basis of the symbiosis between plants and AMF (Panneerselvam et al., 2017, 2019). In a recent study, the expression of the gene RAM1 (Required for Arbuscular Mycorrhization 1) has confirmed that the plant is providing photosynthates to the AFM to establish a symbiosis (Luginbuehl and Oldroyd, 2017). Based on this AMF, other studies (Bao et al., 2019) explored the role of P transporters encoded by two genes, *OsPT11* and *GintPT*, that under different flooding regimes went through a differential expression but by guaranteeing higher levels of P in inoculated plants.

It is important to mention that the role of AMF is not only limited by abiotic conditions but also influenced by fungal diversity and host specificity. The inoculation of rice with *R. irregularis* (Campo et al., 2020) can result in an optimal growth response for some varieties. For others, the nature of the relationship can be completely different, either neutral or negative. This attitude can be attributed to the different genotypes of the single varieties.

The AMFb perform N_2-fixation, nutrient mobilization, and nutrient uptake facilitation to plant hormone, antibiotic and siderophore production, systemic resistance induction, growth factors production, and detoxification of antagonistic substances (Ujvári et al., 2021).

22.4 AMF AND AMF-ASSOCIATED BACTERIA ROLE IN PLANT STRESS TOLERANCE

AMF and AMFb can act as biostimulants, bio-enhancers, and biological control agents exhibiting various PGP traits under diverse environmental conditions (Agnolucci, Turrini, and Giovannetti, 2019). Various recent reports highlighted the role of AMF and AMFb in plants' growth and nutrition and the alleviation of different stress (Agnolucci, Turrini, and Giovannetti, 2019; Mitra et al., 2022). Most of the works on AMF deal with the application of *Acaulospora, Gigaspora*, and *Glomus* genera (Panneerselvam et al., 2017). Few reports describe the role of AMF and AMFb in biotic and abiotic stress tolerance. Nevertheless, as shown in Table 22.1, some recent reports describe the use of AMF in Poaceae family, to which *Oryza* genus belongs. Findings so far collected show that AMF, mainly *Rhizophagus*, induce plant tolerance against salinity, waterlogging, drought, temperature, and heavy metal stresses. MH and PGP bacteria belong to the large group of beneficial bacteria with a major role in the tolerance of many biotic and abiotic stresses. This bacterial ability mainly occurs thanks to the production of several molecules (e.g., elicitors and antibiotics) and the induction of an induced systemic resistance (ISR) in plants (Ujvári et al., 2021).

22.5 CONCLUSION AND FUTURE PERSPECTIVES

The use of AMF for seed priming has several promising potentials for seed technology and agricultural productivity. The studies presented in this chapter show the positive effects of priming on seed quality, seedling growth, and rice productivity, even under stressful conditions. Priming initiates specific signaling pathways in the early stages of plant phenology and leads to a faster plant defense response. In this way, future exposure to pathogens activates a second signaling

TABLE 22.1
Alleviation of Abiotic and Biotic Stress by AMF

Microorganism	Host Plant	Type of Stress	Effects	References
Rhizoglomus intraradices	Rice (*Oryza rufipogon*)	*Magnaporthe oryzae*	Regulation of phytohormones production, plant resistance	Tian et al. (2019)
Claroideoglomus etunicatum	Maize (*Zea mays*)	Cadmium, Lanthanum	Plant growth and resistance to Cd and La stresses, stability of the rhizospheric microbial community	Hao et al. (2022)
Rhizophagus irregularis	Maize (*Zea mays* L.)	Salinity	Plant growth	Chen et al. (2022)
Claroideoglomus etunicatum, Funneliformis mosseae, Gigaspora margarita, Scutellospora calospora	Sewan grass (*Lasiurus scindicus*)	Salinity	Plant growth increased photosynthesis rate and antioxidant activity	Malik et al. (2022)
Glomus spp.	False wheatgrass (*Leymus chinensis*), Limpograss (*Hemarthria altissima*)	Drought	Enhanced plant biomass, antioxidant activity, and photosynthetic rate	Li et al. (2019)
Rhizophagus aggregatus, Funneliformis mosseae	Millet lines (*Panicum miliaceum*)	Temperature	Plant growth, tolerance to temperature stress	Ndeko et al. (2022)
Rhizophagus intraradices, Funneliformis mosseae, Funneliformis geosporum	Wheat (*Triticum aestivum*)	Drought	Increasing RWC and protecting photosynthetic apparatus	Mathur, Tomar, and Jajoo (2019)

pathway that ultimately triggers and amplifies signal transduction, leading to a rapid and more intense activation of the previously acquired defense response. In addition, seed priming has numerous advantages, such as cost efficiency, speed, and environmental friendliness, as well as desirable characteristics of the treated seed compared to chemical treatments. Priming with AMF and AMFb also showed potentiality as tolerance inducers against biotic and abiotic stresses.

Currently, few studies are looking at the combined use of AMF and AMFb in biopriming. Additional research on the subject could provide new insights on an emerging sustainable tool to avoid the problem of reduced rice productivity under abiotic and biotic stresses. Future research must be based on producing a microbial formulation adapted to local ecological conditions. In this direction, using indigenous strains of microorganisms under local conditions should be explored to avoid the risk of introducing unsuitable non-indigenous microbes to local sites.

REFERENCES

Agnolucci, M., A. Turrini, and M. Giovannetti. 2019. Molecular and Functional Characterization of Beneficial Bacteria Associated with AMF Spores. In *Methods in Rhizosphere Biology Research*, eds. D. Reinhardt and A.K. Sharma, pp. 61–79. Singapore: Springer. http://link.springer.com/10.1007/978-981-13-5767-1_5.

Arif, I., M. Batool, and P.M. Schenk. 2020. Plant Microbiome Engineering: Expected Benefits for Improved Crop Growth and Resilience. *Trends in Biotechnology* 38(12): 1385–1396. https://linkinghub.elsevier.com/retrieve/pii/S0167779920301219.

Bao, X. Z., Y. T. Wang, and P. A. Olsson. 2019. Arbuscular mycorrhiza under water-Carbon-phosphorus exchange between rice and arbuscular mycorrhizal fungi under different flooding regimes. *Soil Biology and Biochemistry* 129: 169–177. doi: 10.1016/j.soilbio.2018.11.020

Behrooz, A., K. Vahdati, F. Rejali, M. Lotfi, S. Sarikhani, and C. Leslie. 2019. Arbuscular Mycorrhiza and Plant Growth-Promoting Bacteria Alleviate Drought Stress in Walnut. *HortScience* 54(6): 1087–1092. https://journals.ashs.org/view/journals/hortsci/54/6/article-p1087.xml.

Bernaola, L., G. Cange, M.O. Way, J. Gore, J. Hardke, and M. Stout. 2018. Natural Colonization of Rice by Arbuscular Mycorrhizal Fungi in Different Production Areas. *Rice Science* 25(3): 169–174. https://linkinghub.elsevier.com/retrieve/pii/S1672630818300246.

Campo, S., H. Martín-Cardoso, M. Olivé, E. Pla, M. Catala-Forner, M. Martínez-Eixarch, and B. San Segundo. 2020. Effect of Root Colonization by Arbuscular Mycorrhizal Fungi on Growth, Productivity and Blast Resistance in Rice. *Rice* 13(1): 42. https://thericejournal.springeropen.com/articles/10.1186/s12284-020-00402-7.

Chen, Q., X. Deng, J.T.M. Elzenga, and J.D. van Elsas. 2022. Effect of Soil Bacteriomes on Mycorrhizal Colonization by Rhizophagus Irregularis: Interactive Effects on Maize (*Zea Mays* L.) Growth under Salt Stress. *Biology and Fertility of Soils* 58(5): 515–525. https://link.springer.com/article/10.1007/s00374-022-01636-x.

do Espirito Santo Pereira, A., H. Caixeta Oliveira, L. Fernandes Fraceto, and C. Santaella. 2021. Nanotechnology Potential in Seed Priming for Sustainable Agriculture. *Nanomaterials* 11(2): 267. https://www.mdpi.com/2079-4991/11/2/267.

Giovannini, L., M. Palla, M. Agnolucci, L. Avio, C. Sbrana, A. Turrini, and M. Giovannetti. 2020. Arbuscular Mycorrhizal Fungi and Associated Microbiota as Plant Biostimulants: Research Strategies for the Selection of the Best Performing Inocula. *Agronomy* 10(1): 106. https://www.mdpi.com/2073-4395/10/1/106.

Hao, B., Z. Zhang, Z. Bao, L. Hao, F. Diao, F.Y. Li, and W. Guo. 2022. Claroideoglomus Etunicatum Affects the Structural and Functional Genes of the Rhizosphere Microbial Community to Help Maize Resist Cd and La Stresses. *Environmental Pollution* 307: 119559. https://linkinghub.elsevier.com/retrieve/pii/S0269749122007734.

Li, J., B. Meng, H. Chai, X. Yang, W. Song, S. Li, A. Lu, T. Zhang, and W. Sun. 2019. Arbuscular Mycorrhizal Fungi Alleviate Drought Stress in C3 (Leymus Chinensis) and C4 (Hemarthria Altissima) Grasses via Altering Antioxidant Enzyme Activities and Photosynthesis. *Frontiers in Plant Science* 10: 499. https://www.frontiersin.org/article/10.3389/fpls.2019.00499/full.

Luginbuehl, L.H., and G.E.D. Oldroyd. 2017. Understanding the Arbuscule at the Heart of Endomycorrhizal Symbioses in Plants. *Current Biology* 27(17): R952–R963. https://linkinghub.elsevier.com/retrieve/pii/S0960982217307790.

Luginbuehl, L.H., G.N. Menard, S. Kurup, H. van Erp, G.V. Radhakrishnan, A. Breakspear, G.E.D. Oldroyd, and P.J. Eastmond. 2017. Fatty Acids in Arbuscular Mycorrhizal Fungi Are Synthesized by the Host Plant. *Science* 356(6343): 1175–1178. https://www.science.org/doi/10.1126/science.aan0081.

Majeed, A., Z. Muhammad, and H. Ahmad. 2018. Plant Growth Promoting Bacteria: Role in Soil Improvement, Abiotic and Biotic Stress Management of Crops. *Plant Cell Reports* 37(12): 1599–1609. http://link.springer.com/10.1007/s00299-018-2341-2.

Malik, J.A., A.A. AlQarawi, B.A. Dar, A. Hashem, T.S. Alshahrani, M.N. AlZain, M.M. Habib, M.M. Javed, and E.F. Abd_Allah. 2022. Arbuscular Mycorrhizal Fungi Isolated from Highly Saline "Sabkha Habitat" Soil Alleviated the NaCl-Induced Stress and Improved Lasiurus Scindicus Henr. Growth. *Agriculture* 12(3): 337. https://www.mdpi.com/2077-0472/12/3/337.

Mathur, S., R.S. Tomar, and A. Jajoo. 2019. Arbuscular Mycorrhizal Fungi (AMF) Protects Photosynthetic Apparatus of Wheat under Drought Stress. *Photosynthesis Research* 139(1–3): 227–238. http://link.springer.com/10.1007/s11120-018-0538-4.

Mitra, D., B. Saritha, E. Janeeshma, P. Gusain, B. Khoshru, F.A. Abo Nouh, A. Rani, et al. 2022. Arbuscular Mycorrhizal Fungal Association Boosted the Arsenic Resistance in Crops with Special Responsiveness to Rice Plant. *Environmental and Experimental Botany* 193: 104681. https://linkinghub.elsevier.com/retrieve/pii/S0098847221003117.

Mitra, D., R. Djebaili, M. Pellegrini, B. Mahakur, A. Sarker, P. Chaudhary, B. Khoshru, et al. 2021. Arbuscular Mycorrhizal Symbiosis: Plant Growth Improvement and Induction of Resistance under Stressful Conditions. *Journal of Plant Nutrition* 44(13): 1993–2028. https://www.tandfonline.com/doi/full/10.1080/01904167.2021.1881552.

Ndeko, A.B., H. Founoune-Mboup, A. Kane, and L. Cournac. 2022. Arbuscular Mycorrhizal Fungi Alleviate the Negative Effect of Temperature Stress in Millet Lines with Contrasting Soil Aggregation Potential. *Gesunde Pflanzen* 74(1): 53–67. https://link.springer.com/10.1007/s10343-021-00588-w.

Panneerselvam, P., S. Sahoo, A. Senapati, U. Kumar, D. Mitra, C. Parameswaran, A. Anandan, A. Kumar, A. Jahan, and A.K. Nayak. 2019. Understanding Interaction Effect of Arbuscular Mycorrhizal Fungi in Rice under Elevated Carbon Dioxide Conditions. *Journal of Basic Microbiology* 59(12): 1217–1228.

Panneerselvam, P., U. Kumar, T.C.K. Sugitha, C. Parameswaran, S. Sahoo, A.K. Binodh, A. Jahan, and A. Anandan. 2017. Arbuscular Mycorrhizal Fungi (AMF) for Sustainable Rice Production. In *Advances in Soil Microbiology: Recent Trends and Future Prospects*, eds. K.A. Tapan, B.M. Bibhuti, K. Annapurna, K.V. Deepak, and K. Upendra, pp. 99–126. Singapore: Springer. http://link.springer.com/10.1007/978-981-10-7380-9_6.

Saha, D., P. Choyal, U.N. Mishra, P. Dey, B. Bose, M.D. Prathibha, N.K. Gupta, et al. 2022. Drought Stress Responses and Inducing Tolerance by Seed Priming Approach in Plants. *Plant Stress* 4: 100066. https://linkinghub.elsevier.com/retrieve/pii/S2667064X22000124.

Sangwan, S., and R. Prasanna. 2022. Mycorrhizae Helper Bacteria: Unlocking Their Potential as Bioenhancers of Plant–Arbuscular Mycorrhizal Fungal Associations. *Microbial Ecology* 84(1): 1–10. https://link.springer.com/10.1007/s00248-021-01831-7.

Sarkar, D., S. Singh, M. Parihar, and A. Rakshit. 2021. Seed Bio-Priming with Microbial Inoculants: A Tailored Approach towards Improved Crop Performance, Nutritional Security, and Agricultural Sustainability for Smallholder Farmers. *Current Research in Environmental Sustainability* 3: 100093. https://linkinghub.elsevier.com/retrieve/pii/S2666049021000694.

Shasmita, B.B. Swain, P.K. Mohapatra, S.K. Naik, and A.K. Mukherjee. 2022. Biopriming for Induction of Disease Resistance against Pathogens in Rice. *Planta* 255(6): 113. https://link.springer.com/10.1007/s00425-022-03900-8.

Tian, L., C. Chang, L. Ma, F. Nasir, J. Zhang, W. Li, L.-S.P. Tran, and C. Tian. 2019. Comparative Study of the Mycorrhizal Root Transcriptomes of Wild and Cultivated Rice in Response to the Pathogen Magnaporthe Oryzae. *Rice* 12(1): 35. https://thericejournal.springeropen.com/articles/10.1186/s12284-019-0287-9.

Ujvári, G., A. Turrini, L. Avio, and M. Agnolucci. 2021. Possible Role of Arbuscular Mycorrhizal Fungi and Associated Bacteria in the Recruitment of Endophytic Bacterial Communities by Plant Roots. *Mycorrhiza* 31(50): 527–544. https://link.springer.com/10.1007/s00572-021-01040-7.

Wagner, K., K. Krause, R. Gallegos-Monterrosa, D. Sammer, Á.T. Kovács, and E. Kothe. 2019. The Ectomycorrhizospheric Habitat of Norway Spruce and Tricholoma Vaccinum: Promotion of Plant Growth and Fitness by a Rich Microorganismic Community. *Frontiers in Microbiology* 10. https://www.frontiersin.org/article/10.3389/fmicb.2019.00307/full.

23 Arbuscular Mycorrhizal Fungi in the Control of Fungal Diseases in Rice

Jorge Poveda
Universidad Pública de Navarra (UPNA)
Instituto Politécnico de Bragança

CONTENTS

23.1 INTRODUCTION

Rice (*Oryza sativa*) is a cereal that has been cultivated for more than 7,000 years and is the fundamental basis of the diet of more than half of the world's population (Panneerselvam et al., 2017; Maiti et al., 2020). In rice cultivation, losses associated with pathogens and/or pests account for 25%–41% of total production (Savary et al., 2019). Regarding pathogens, rice plants can be infected by viruses, bacteria, fungi, oomycetes, and nematodes, causing a total of more than 70 different diseases (Faizal-Azizi and Lau, 2022). The two main strategies used globally for disease control in rice cultivation are the use of resistant varieties and chemical pesticides (Narasimhamurthy et al., 2021). However, the massive use of these agricultural chemicals causes serious environmental and health problems, in addition to affecting rice quality (Teló et al., 2015). Therefore, it is necessary to search for new alternatives to control rice diseases, and arbuscular mycorrhizal fungi (AMF) may be a good strategy. This chapter compiles and analyzes all the work done so far on the control of fungal diseases in rice using AMF.

DOI: 10.1201/9781003354086-23

23.2 AMF AS BIOLOGICAL CONTROL AGENTS

In addition to nutritional and abiotic stress benefits, AMF stand out as efficient biological control agents (BCAs), both against pathogens and pests of agricultural importance (Malik, 2018; Poovarasan et al., 2013; Mohandas et al., 2013). Against plant pathogens, AMF act as BCAs through different direct and indirect mechanisms of action (Wehner et al., 2010; Dowarah et al., 2021), improving the nutritional status of their host plant (Wehner et al., 2010).

Direct mechanisms include those related to competition (Wehner et al., 2010). AMF have very powerful nutrient uptake mechanisms that can cause deficiencies in the rhizosphere and prevent their acquisition by soilborne pathogens (Dowarah et al., 2021). Before penetrating roots, pathogens and AMF should occupy the same rhizospheric space and "enter" through the same infection sites (Dowarah et al., 2021). An indirect mechanism of AMF as BCAs is the active modification of the rhizospheric microbiota that leads to the proliferation of microorganisms antagonistic to plant pathogens (Wehner et al., 2010; Dowarah et al., 2021).

However, the main mechanism of action studied in AMF as BCAs is the local and systemic activation of plant defenses (Wehner et al., 2010; Dowarah et al., 2021). Although plant colonization by AMF is restricted to roots, this interaction activates the so-called mycorrhiza-induced resistance (MIR) (Dowarah et al., 2021).

23.3 FUNGAL DISEASES OF RICE

The rice crop is mainly attacked by eight different fungal species of worldwide distribution. The descriptions of the diseases are compiled in Table 23.1.

23.4 AMFs AS BCAs OF FUNGAL PATHOGENS ON RICE

Although the mechanisms of action are not always known, AMF have been described as an efficient biological control tool against fungal diseases in rice. In the field, root application of *Glomus* sp. reduced the incidence of sheath blight (produced by *R. solani*) by 30%. The AMF were applied at transplanting of rice seedlings, while the pathogen was applied one month later (Ilag et al., 1987). In a work carried out under greenhouse conditions with a formulation of several AMF species (*Glomus fasciculatum*, *G. mosseae*, *G. aggregatum*, *G. fulvum*, *Gigaspora candida*, and *G. gigantea*), it was determined that the mechanism of action involved in the reduction of disease caused by *R. solani* was an active competition for root space and infection sites (Baby and Manibhushanrao, 1996), a mechanism widely described for AMF such as BCAs.

The main mechanism reported in the control of fungal diseases in rice by AMF is the systemic activation of defenses by BCAs root colonization. In a study conducted under greenhouse conditions with the AMF *Funneliformis mosseae* and *Rhizophagus irregularis*, it was found that the aerial part of the plant was less affected by the pathogen *M. oryzae* (the cause of blast disease). However, the reported reduction of the disease could also be due to an increase in the plant

TABLE 23.1
Main Fungal Diseases of Rice

Fungal Pathogen				
Preferred Scientific Name	**Other Scientific Names**	**Disease**	**Symptoms**	**Distribution**
Cochliobolus miyabeanus	*Bipolaris oryzae* *Drechslera oryzae* *Helminthosporium oryzae* *Ophiobolus miyabeanus*	Brown leaf spot	*Seeds*: lesions on seeds *Stems*: discoloration of bark *Leaves*: abnormal colors, fungal growth *Inflorescence*: lesions on glumes	Widely distributed worldwide
Gibberella fujikuroi	*Alysidium verticillioides* *Fusarium moniliforme* *F. neoceras* *F. sacchari* *F. subglutinans* *F. verticillioides* *Gibberella moniliforme* *Lisea fujikuroi* *Oospora verticillioides*	Bakanae disease	*Whole plant*: damping off, early senescence, seedling blight *Seeds*: distortion, lesions on seeds, rot *Roots*: soft rot of cortex *Stems*: discoloration of bark, mold growth on lesion, stunting, or resetting *Leaves*: abnormal colors, fungal growth *Fruit*: lesions black or brown, lesions scab or pitting	Widely distributed worldwide
Magnaporthe oryzae	*Dactylaria oryzae* *Piricularia oryzae* *Pyricularia oryzae*	Blast disease	*Whole plant*: damping off, plant dead, dieback *Seeds*: lesions on seeds, rot *Stems*: internal red necrosis, mold growth on lesion, stunting or resetting *Leaves*: abnormal colors, abnormal forms, abnormal leaf fall, fungal growth, necrotic areas, rot *Inflorescence*: discoloration panicle, lesions on glumes, rot	Widely distributed worldwide

(*Continued*)

TABLE 23.1 ***Continued***
Main Fungal Diseases of Rice

Fungal Pathogen: Preferred Scientific Name	Fungal Pathogen: Other Scientific Names	Disease	Symptoms	Distribution
Magnaporthe salvinii	*Curvularia sigmoidea* *Helminthosporium sigmoideum* *Leptosphaeria salvinii* *Nakataea sigmoidea* *Sclerotium oryzae* *Vakrabeeja sigmoidea*	Stem rot	*Seeds*: lesions on seeds, rot *Stems*: discoloration of bark, mold growth on lesion, mycelium present. *Leaves*: fungal growth, necrotic areas, rot *Inflorescence*: discoloration panicle, lesions on glumes	North America, South America, South Europe, Asia and Oceania
Monographella albescens	*Gerlachia oryzae* *Griphosphaerella albescens* *Metasphaeria albescens* *M. oryzae-sativae* *Miconectriella pavgii* *Microdochium oryzae* *Rhynchosporium oryzae*	Leaf scald	*Whole plant*: seedling blight *Seeds*: rot *Roots*: soft rot of cortex *Leaves*: abnormal colors, fungal growth, necrotic areas *Inflorescence*: discoloration panicle, lesions on glumes	Asia, North America, South America and Oceania
Rhizoctonia solani	*Thanatephorus cucumeris*	Sheath blight	*Leaves*: necrotic areas, rot	Widely distributed worldwide

(Continued)

TABLE 23.1 ***Continued***
Main Fungal Diseases of Rice

Fungal Pathogen				
Preferred Scientific Name	**Other Scientific Names**	**Disease**	**Symptoms**	**Distribution**
Sarocladium oryzae	*Acrocylindrium oryzae* *Sarocladium attenuatum*	Sheath rot	*Seeds*: discolorations, rot *Leaves*: abnormal colors, fungal growth *Inflorescence*: discoloration panicle, lesions on glumes, twisting and distortion	Asia, North America, South America and Oceania
Tilletia barclayana	*Neovossia barclayana* *N. horrida* *Tilletia ajrekari* *T. horrida* *T. pulcherrima*	Black smut	*Seeds*: discolorations, distortion, lesions on seeds *Vegetative organs*: surface lesions or discoloration *Inflorescence*: black fungal spores	North America, South America, Asia and Oceania

nutritional content, since a significant increase in the foliar content of inorganic P was quantified in AMF-inoculated plants (Campo et al., 2020).

In two studies conducted under growth chamber conditions, activation of systemic resistance against *M. oryzae* could be reported by transcriptomic and metabolomic analyses. Root inoculation of rice plants with *Glomus intraradices* reduced blast disease dryness through a systemic increase in the expression of the defense-related genes *OsNPR1*, *OsAP2*, *OsEREBP*, and *OsJAmyb*. Additionally, an increase in the expression of genes involved in signal transduction processes (*OsDUF26* and *OsMPK6*) and in calcium-mediated signaling processes (*OsCBP*, *OsCaM* and *OsCML4*) was also reported (Campos-Soriano et al., 2012). Using the same AMF (renamed *Rhizoglomus intraradices*), it was determined how this increased systemic resistance in rice leaves against *M. oryzae* was involved in accumulating a series of defense-related compounds. These secondary metabolites are within the biosynthetic pathways of α-linolenic acid, phenols, terpenoids, and mevalonate (Tian et al., 2019).

Greenhouse and field trials reported reductions of 45% and 65%, respectively, in rice brown spot disease (caused by *B. oryzae*) due to AMF root colonization. This disease biocontrol action is a consequence of the systemic activation of rice defenses by AMF, through an increase in leaf activity of the enzyme-defense-related catalase, peroxidase, and polyphenol oxidase (Saleh and El-Akshar, 2020).

Although root colonization of rice by AMF can increase systemic plant defenses, it requires a decrease in local root defenses at some point to allow colonization by the beneficial fungus. In this sense, it was found that rice plants root inoculated with the AMF *R. irregularis*, *R. fasciculatum*, *F. mosseae*, *Glomus deserticola*, *G. microaggregatum*, and *Sclerocystis dussii* were more susceptible to sheath blight disease. This was determined to be a consequence of decreased local root defenses by AMF colonization, which allowed *R. solani* to more successfully colonize rice roots (Bernaola et al., 2018).

A special case is that of a mycorrhizal fungus isolated from orchid roots, which can be cultivated in artificial growth media, although it does not colonize rice roots, the species *Waitea circinata*. Numerous studies have been conducted with this fungus as BCAs in different crops, including rice, through the application of cell-free filters. Both *in vitro* and in greenhouse trials, it has been reported that these filtrates are capable of reducing conidia germination (up to 85%), mycelium growth (up to 75%), and disease incidence (up to 90%) of the rice pathogens *C. miyabeanus*, *M. oryzae*, *Monographella albescens,* and S. *oryzae* (Carvalho et al., 2015, 2021, 2022). This biocontrol capacity is due to the production of various antifungal tools by *W. circinata*, which includes the production of diffusible secondary metabolites, such as benzophenones, and lytic enzymes, such as glucanases, chitinases, and proteases (Carvalho et al., 2015, 2021, 2022).

23.5 CONCLUSION

Thus far, very few studies have been developed using AMF as BCAs against fungal diseases of rice. These studies have reported important results on disease control efficiency. Alternatively, more research is needed: on the one hand, through field studies that allow knowing their real efficiency in the agricultural system, and on the other hand, through the use of new analysis tools within the field of OMICS, being able to identify the exact mechanisms involved in the capacity of AMFs as BCAs of these diseases. Transcriptomic, proteomic, hormonomic, and/or metabolomic techniques will greatly expand the knowledge on the biology of AMF in their symbiosis with rice and against fungal pathogens attack. Therefore, AMF represent an efficient tool in the control of fungal diseases of the rice crop. However, much research is needed to establish their use in the global agricultural system.

ACKNOWLEDGMENTS

Grants for the Recualification of the Spanish University System for 2021–2023, Public University of Navarra; Recualification Modality; Funded by the European Union – NextGenerationEU.

REFERENCES

Baby, U. I., and K. Manibhushanrao. 1996. Influence of organic amendments on arbuscular mycorrhizal fungi in relation to rice sheath blight disease. *Mycorrhiza* 6:201–206.

Bernaola, L., Cosme, M., Schneider, R. W., and M. Stout. 2018. Belowground inoculation with arbuscular mycorrhizal fungi increases local and systemic susceptibility of rice plants to different pest organisms. *Frontiers in Plant Science* 9:747.

Campo, S., Martín-Cardoso, H., Olivé, M., Pla, E., Catala-Forner, M., Martínez-Eixarch, M., and B. San Segundo. 2020. Effect of root colonization by arbuscular mycorrhizal fungi on growth, productivity and blast resistance in rice. *Rice* 13:1–14.

Campos-Soriano, L. I. D. I. A., García-Martínez, J., and B. S. Segundo. 2012. The arbuscular mycorrhizal symbiosis promotes the systemic induction of regulatory defence-related genes in rice leaves and confers resistance to pathogen infection. *Molecular Plant Pathology* 13:579–592.

Carvalho, J., Chaibub, A., Sousa, K., Brito, D., Silva, C., Moller, P., et al. 2022. Characterization of the antagonism of the mycorrhizal fungus Waitea circinata against Magnaporthe oryzae, Cochliobolus miyabeanus, Monographella albescens and *Sarocladium oryzae* rice pathogens. Colloquium Agrariae 18:1–14.

Carvalho, J. C. B. D., Chaibub, A. A., Sousa, K. C. I., Brito, D. C. D., Filippi, M. C. C. D., Kato, L., and L. G. D. Araújo. 2021. Efficiency of a new Waitea circinata extract against rice pathogens. Pesquisa Agropecuária Tropical 51:1–9.

Carvalho, J. C., Sousa, K. C., Brito, D. C., Chaibub, A. A., Luzini, A. P., Côrtes, M. V., et al. 2015. Biocontrol potential of *Waitea circinata*, an orchid mycorrhizal fungus, against the rice blast fungus. *Tropical Plant Pathology* 40:151–159.

Dowarah, B., Gill, S. S., and N. Agarwala. 2021. Arbuscular mycorrhizal fungi in conferring tolerance to biotic stresses in plants. *Journal of Plant Growth Regulation* 41:1429–1444.

Faizal-Azizi, M. M., and H. Y. Lau. 2022. Advanced diagnostic approaches developed for the global menace of rice diseases: a review. *Canadian Journal of Plant Pathology*. https://doi.org/10.1080/07060661.2022.2053588.

Ilag, L. L., Rosales, A. M., and T. W. Mew. 1987. Vesicular-arbuscular fungus reduces Rhizoctonia disease incidence. In Asia-Pacific Biotechnology Congress, Manila, Philippines, 11–14 October 1986. Philippine Society of Microbiology, Inc.

Maiti, R., Sarkar, N. C., Rodriguez, H. G., Kumari, A., Begum, S., and D. Rajkumar. 2020. Advances in Rice Science: Botany, Production, and Crop Improvement. Apple Academic Press: Pleasant, NJ.

Malik, R. J. 2018. Recent trend: is the role of arbuscular mycorrhizal fungi in plant-enemies performance biased by taxon usage? *The American Midland Naturalist* 180:306–311.

Mohandas, S., Poovarasan, S., Panneerselvam, P., Saritha, B., Upreti, K. K., Kamal, R. and T. Sita. 2013. Guava (*Psidium guajava* L.) rhizosphere Glomus mosseae spores harbor actinomycetes with growth promoting and antifungal attributes. *Scientia Horticulturae* 150:371–376.

Narasimhamurthy, H. B., Naik, G., Sehgal, M., and M. Malik. 2021. Integrated management of rice diseases. *Journal of Advanced Research in Agriculture Science and Technology* 4:13–24.

Panneerselvam, P., Kumar, U., Sugitha, T. C. K., Parameswaran, C., Sahoo, S., Binodh, A. K., Jahan, A. and A. Anandan. 2017. Arbuscular mycorrhizal fungi (AMF) for sustainable rice production. In: T. K. Adhya, B. Lal, B. Mohapatra, D. Paul, and S. Das (Eds.), *Advances in Soil Microbiology: Recent Trends and Future Prospects* (pp. 99–126). Springer: Singapore.

Poovarasan, S., Mohandas, S., Paneerselvam, P., Saritha, B. and K. M. Ajay. 2013. Mycorrhizae colonizing actinomycetes promote plant growth and control bacterial blight disease of pomegranate (*Punica granatum* L. cv Bhagwa). *Crop Protection* 53:175–181.

Saleh, M. M., and Y. S. El-Akshar. 2020. Integration between arbuscular mycorrhizal fungi, bacterial and fungal bioagents for controlling rice brown spot disease. *Egyptian Journal of Phytopathology* 48:81–93.

Savary, S., Willocquet, L., Pethybridge, S. J., Esker, P., McRoberts, N., and A. Nelson. 2019. The global burden of pathogens and pests on major food crops. *Nature Ecology and Evolution* 3:430–439.

Teló, G. M., Marchesan, E., Zanella, R., Limberger de Oliveira, M., Coelho, L. L., and M. L. Martins. 2015. Residues of fungicides and insecticides in rice field. *Agronomy Journal* 107:851–863.

Tian, L., Chang, C., Ma, L., Nasir, F., Zhang, J., Li, W., et al. 2019. Comparative study of the mycorrhizal root transcriptomes of wild and cultivated rice in response to the pathogen *Magnaporthe oryzae*. *Rice* 12:1–19.

Wehner, J., Antunes, P. M., Powell, J. R., Mazukatow, J., and M. C. Rillig. 2010. Plant pathogen protection by arbuscular mycorrhizas: a role for fungal diversity? *Pedobiologia* 53:197–201.

24 Arbuscular Mycorrhizal Fungi Role in Bioremediation in Rice in the Context of Climate Change

Sarah González Henao and
Thaura Ghneim-Herrera
Universidad ICESI

CONTENTS

24.1 HEAVY METAL POLLUTION IN THE RICE ECOSYSTEM

Rice is one of the main crops for human consumption widely exposed to toxic heavy metals (HMs) in soils owing to wastewater irrigation, and the use of fertilizers and pesticides containing metals have led to the accumulation of HMs in soils (Zulkafflee et al., 2019). High levels of HMs in soils, such as arsenic and cadmium, constitute a major concern to food security, the environment, and the food chain since they are widespread, biologically toxic, and persistent in soils (Zhao et al., 2022).

Climate change is a detrimental factor to rice production as most abiotic stresses are likely to increase due to high temperatures, drought, salinity, and flooding that lead to water scarcity, insects and pests, diseases, and weeds (Mahdu, 2019). Global warming predicting models and *in situ* experiments on

DOI: 10.1201/9781003354086-24

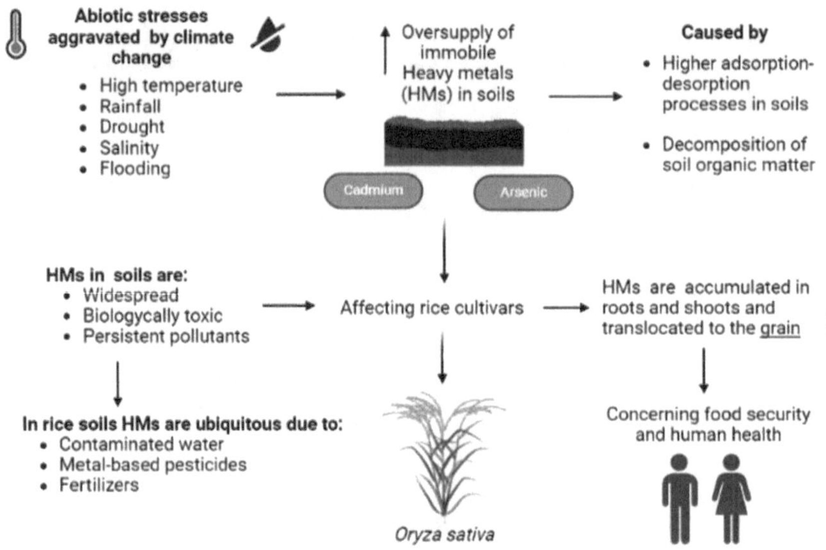

FIGURE 24.1 Overview of the impacts of heavy metals in rice cultivated soils under a climate change scenario.

rice manifest the requirement of rice genotypes with higher tolerance to heat and elevated concentrations of CO_2 to face future climate change and the increasing demand in different countries (Pickson et al., 2022). In Figure 24.1, we show an overview of the consequences of climate change on rice in soils contaminated with HMs.

24.2 HEAVY METALS ACCUMULATION IN RICE ORGANS AND TRANSLOCATION TO THE GRAIN

Arsenic (As) and cadmium (Cd) are two toxic elements that have a high risk of transfer to rice grains, a major source of these two elements for populations consuming rice as a staple food (Zhao and Wang, 2020).

Arsenic (As) is a widespread toxic metalloid in soils which exists in various inorganic and organic forms in the environment; the most prevalent inorganic forms are arsenate (As^{5+}), which is abundant under aerobic conditions, and arsenite (As^{3+}), which is found in anaerobic conditions (Awasthi et al., 2017). In paddy soils (anaerobic conditions), rice plants can accumulate As at several-fold higher levels than other cereal crops such as wheat and maize; for instance, As levels over 0.60 mg/kg were found in rice grains from plants grown in paddy soils containing ~10 mg/kg As (Williams et al., 2007).

On the contrary, cadmium (Cd) constitutes a toxic metal for human health, whose presence in soils is accelerating due to industrialization and intensified agriculture (He et al., 2015). Cd accumulation in rice grains is common but also

TABLE 24.1
Maximum Arsenic and Cadmium Levels in Soils and Rice Grains

Parameter	Inorganic Arsenic (mg/kg)	References	Cadmium (mg/kg)	References
Maximum level reported in soils	11.37	Islam et al. (2004)	8.77	Ji et al. (2012)
Maximum allowable limit (MAL) in soils	20	Kabata-Pendias (2010)	1.0	Ji et al. (2012)
Maximum level reported on rice grains	2.05	Islam et al. (2004)	0.96	Ji et al. (2012)
Maximum concentrations recommended on rice grains	0.2	Asakura and Wu (2015)	0.2	World Health Organization (2005)

variable depending on genotypes; Song et al. (2015) showed a significant Cd concentration (0.14–0.81 mg/kg) in rice grains from 20 rice cultivars exposed to increasing Cd content in the soil (0–0.6 mg/kg). To outline the concern in terms of food safety and human health, in Table 24.1, we compare the maximum reported concentrations of arsenic and cadmium in rice cultivated soils and in rice grains, vs. their maximum allowed concentrations.

24.3 ROLE OF AMF FOR HEAVY METAL TOLERANCE IN RICE

Arbuscular mycorrhizal fungi (AMF) are ubiquitous rhizosphere fungi able to establish symbiotic interactions with plant roots that constitute a potential tool capable of alleviating metal toxicity in plants by reducing metal translocation from root to shoot (Mishra et al., 2016; Wu et al., 2016). In the following section, we discuss the mechanisms by which AMF exert their protective role in soils against arsenic and cadmium pollution.

24.3.1 Arsenic

Symbiosis with AMF improves As tolerance in rice plants, reduces As uptake from the soil, and stimulates many traits related to stress tolerance in rice including activities of antioxidant enzymes, greater biomass accumulation and relative chlorophyll content, increased water-use efficiency, higher carbon assimilation rate, stomatal conductance, and transpiration rates (de Andrade et al., 2015; Pathare et al., 2017).

In rice plants, As^{5+} and phosphorus (P), and As^{3+} and silicon (Si) share the same transporters: As^{5+} and P transport. They involve the Pht1 family genes (*OsPT1–OsPT13*) (Ye et al., 2017), whereas *OsLsi1* and *OsLsi2* are genes associated with

As^{3+} and Si mobilization that encode transporters of the Nodulin-26-like intrinsic proteins (NIP) subfamily of aquaporins in rice roots (de Andrade et al., 2015). It has been observed that a differential expression of these genes encodes for As^{5+} and As^{3+} transporters in AMF-inoculated rice plants in arsenic-polluted soils; for instance, Chen et al. (2013) revealed an increased expression of P transporters (*OsPT1–13*) that led to a significantly higher P concentration in plant tissues and consequently lowered As ratio uptake in mycorrhizal rice plants treated with As^{5+}. Moreover, a significant decrease in transporters *OSLsi1* and *OsLsi2* was reported by Chen et al. (2012), which led to a significant reduction of As^{3+} uptake in rice.

24.3.2 Cadmium

Several studies have shown a positive effect of AMF in alleviating cadmium pollution in rice by decreasing significantly Cd concentrations in shoots and roots (Chen et al., 2019; Li et al., 2020; Gao et al., 2021). Inoculation of AMF in rice exposed to cadmium has shown to improve many traits of the plant: elevated chlorophyll contents, improved mesophyll cell structure, decreased lipid peroxidation and detoxification of free radicals, promoted rice growth, enhanced shoot biomass, plant height and root length, and enhanced phosphorus acquisition (Li et al., 2020; Gao et al., 2021; Zhu et al., 2022).

The immobilization of Cd in the cell wall and vacuoles play an important role in Cd detoxification (Gao et al., 2021). Li et al. (2016) found that Cd concentrations were reduced by AMF in shoots and roots; at low Cd levels (<0.05 mM), the cell wall was involved in Cd retention, while at high Cd concentrations (≥ 0.05 mM), vacuoles contributed to Cd detoxification. In another study, Gao et al. (2021) showed that Cd in rice roots inoculated with AMF was mainly found in pectin and hemicellulose 1 (HC1) components of the root cell wall, and inoculation with AMF increased the levels of pectin, HC1, and lignin content, accompanied by increments of L-phenylalanine ammonia-lyase (PAL) and pectin methylesterase (PME) activities.

At a molecular level, inoculation of rice with AMF reduced rice Cd uptake by changing the expression of Cd transporters (Li et al., 2020; Zhu et al., 2022). For instance, Zhu et al. (2022), identified 229 differentially expressed genes (DEGs) in upland rice root inoculated with *Glomus versiforme*; the gene *Nramp5* involved in manganese (Mn) and Cd transport was significantly downregulated, whereas genes related to cysteine biosynthesis, metalloprotein, peroxidase, peroxisome, and carotenoid synthesis were upregulated and elucidated the molecular mechanism by which AMF reduce the accumulation and toxicity of Cd in upland rice.

24.4 VIABILITY OF USING AMF IN RICE UNDER HEAVY METAL SOILS

The selection of appropriate conditions to harvest rice under HM polluted soils must be considered for food security purposes. It has been shown that AMF is not suitable for growing under anaerobic or flooded conditions since colonization rates tend to decrease due to flooded conditions on rice cultivars under HM pollution, and the proportion of metalloids such as arsenic can increase in grain (Li et al., 2013; Zhang et al., 2016). Furthermore, the selection of an appropriate combination of AMF and genotype of rice cultivar under polluted soils is desirable to reduce HM uptake in rice grains, as several studies focused on arsenic and cadmium pollution have reported their translocation to the grain and a differential accumulation of these metals in rice organs depending on the genotype (Zhang et al., 2016; Ueno et al., 2009).

24.5 CONCLUSION

AMF constitute a potential tool that can be extensively used to reduce HMs uptake by rice plants and prevent their accumulation in rice organs and translocation to the grain under a climate change context, a scenario where HM concentrations in soils could be dramatically increased. Future works should focus on investigating the best combination of AMF and rice genotypes that maximize the potential of AMF and therefore prevent HM absorption and accumulation in rice such as As and Cd.

REFERENCES

Asakura, K. and Y. Wu. 2015. Codex committee on contaminants in foods: proposed draft code of practice for the prevention and reduction of arsenic contamination in rice. In *Joint FAO/WHO Food Standards Programme Codex Committee on Contaminants in Foods*, New Delhi, India.

Awasthi, S., R. Chauhan, S. Srivastava and R. D. Tripathi. 2017. The journey of arsenic from soil to grain in rice. *Frontiers in Plant Science* 8: 1007.

Chen, X., F. Wu, H. Li et al. 2013. Phosphate transporters expression in rice (*Oryza sativa* L.) associated with arbuscular mycorrhizal fungi (AMF) colonization under different levels of arsenate stress. Environmental and *E*xperimental *B*otany 87: 92–99.

Chen, X., H. Li, W. F. Chan et al. 2012. Arsenite transporters expression in rice (Oryza sativa L.) associated with arbuscular mycorrhizal fungi (AMF) colonization under different levels of arsenite stress. *Chemosphere* 89(10): 1248–1254.

Chen, X. W., L. Wu, N. Luo, C. H. Mo, M. H. Wong and H. Li. 2019. Arbuscular mycorrhizal fungi and the associated bacterial community influence the uptake of cadmium in rice. *Geoderma* 337: 749–757.

de Andrade, S. A. L., A. P. Domingues Jr and P. Mazzafera. 2015. Photosynthesis is induced in rice plants that associate with arbuscular mycorrhizal fungi and are grown under arsenate and arsenite stress. Chemosfere 134: 141–149.

Gao, M. Y., X. W. Chen, W. X. Huang et al. 2021. Cell wall modification induced by an arbuscular mycorrhizal fungus enhanced cadmium fixation in rice root. *Journal of Hazadous Materials* 416: 125894.

He, S., Z. He, X. Yang, P. J. Stoffella and V. C. Baligar. 2015. Soil biogeochemistry, plant physiology, and phytoremediation of cadmium-contaminated soils. Advances in Agronomy 134: 135–225.

Islam, M., M. Jahiruddin, and S. Islam. 2004. Assessment of arsenic in the water-soil-plant systems in gangetic flood plains of Bangladesh. *Asian Journal of Plant Sciences (Pakistan)* 3: 489–493.

Ji, W., Z. Chen, D. Li and W. Ni. 2012. Identifying the criteria of cadmium pollution in paddy soils based on a field survey. Energy Procedia 16: 27–31.

Kabata-Pendias, A. 2010. *Trace Elements in Soils and Plants*. Boca Raton, FL: CRC Press.

Li, H., N. Luo, L. J. Zhang et al. 2016. Do arbuscular mycorrhizal fungi affect cadmium uptake kinetics, subcellular distribution and chemical forms in rice? Science of the Total Environment 571: 1183–1190.

Li, H., X. W. Chen, L. Wu et al. 2020. Effects of arbuscular mycorrhizal fungi on redox homeostasis of rice under Cd stress. *Plant and Soil* 455(1): 121–138.

Li, H., Y. B. Man, Z. Ye, C. Wu, S. Wu and M. H. Wong. 2013. Do arbuscular mycorrhizal fungi affect arsenic accumulation and speciation in rice with different radial oxygen loss? *Journal of H*azardous *M*aterials 262: 1098–1104.

Mahdu, O. 2019. The Impacts of Climate Change on Rice Production and Small Farmers' Adaptation: A Case of Guyana. PhD dissertation, Virginia Tech.

Mishra, V., A. Gupta, P. Kaur et al. 2016. Synergistic effects of arbuscular mycorrhizal fungi and plant growth promoting rhizobacteria in bioremediation of iron contaminated soils. *International Journal of Phytoremediation* 18(7): 697–703.

Pathare, V., A. Shukla, S. Srivastava and P. Suprasanna. 2017. Response of rice-mycorrhizal association to arsenate exposure. *Bulletin of Environmental and Scientific Research* 6(2): 1–6.

Pickson, R. B., G. He and E. Boateng. 2022. Impacts of climate change on rice production: evidence from 30 Chinese provinces. *Environment, Development and Sustainability* 24(3): 3907–3925.

Song, W.-E., S.-B. Chen, J.-F. Liu et al. 2015. Variation of Cd concentration in various rice cultivars and derivation of cadmium toxicity thresholds for paddy soil by species-sensitivity distribution. *Journal of Integrative Agriculture* 14(9): 1845–1854.

Ueno, D., I. Kono, K. Yokosho, T. Ando, M. Yano and J. F. Ma. 2009. A major quantitative trait locus controlling cadmium translocation in rice (Oryza sativa). *New Phytologist* 182 (3): 644–653.

Williams, P.N., A. Villada, C. Deacon, A. Raab, J. Figuerola, A.J. Green, J. Feldmann, and A.A. Meharg. 2007. Greatly enhanced arsenic shoot assimilation in rice leads to elevated grain levels compared to wheat and barley. *Environmental Science & Technology*, 41(19): 6854–6859.

World Health Organization. 2005. Report of the Twenty-Eighth Session of the Joint FAO/WHO Codex Alimentarius Commission. https://www.fao.org/3/a0214e/a0214e.pdf.

Wu, S., X. Zhang, B. Chen et al. 2016. Chromium immobilization by extraradical mycelium of *Arbuscular mycorrhiza* contributes to plant chromium tolerance. Environmental and Experimental Botany 122: 10–18.

Ye, Y., P. Li, T. Xu et al. 2017. OsPT4 contributes to arsenate uptake and transport in rice. *Frontiers in Plant Science* 8: 2197.

Zhang, X., S. Wu, B. Ren and B. Chen. 2016. Water management, rice varieties and mycorrhizal inoculation influence arsenic concentration and speciation in rice grains. Mycorrhiza 26(4): 299–309.

Zhao, F.-J. and P. Wang. 2020. Arsenic and cadmium accumulation in rice and mitigation strategies. *Plant and Soil* 446(1): 1–21.

Zhao, H., Y. Wu, X. Lan, Y. Yang, X. Wu and L. Du. 2022. Comprehensive assessment of harmful heavy metals in contaminated soil in order to score pollution level. *Scientific Reports* 12(1): 1–13.

Zhu, Q., P. Xu, L. Lei and Y. Jing. 2022. Transcriptome analysis reveals decreased accumulation and toxicity of Cd in upland rice inoculated with arbuscular mycorrhizal fungi. *Applied Soil Ecology* 177: 104501.

Zulkafflee, N. S., N. A. M. Redzuan, Z. Hanafi et al. 2019. Heavy metal in paddy soil and its bioavailability in rice using in vitro digestion model for health risk assessment. *International Journal of Environmental Research and Public Health* 16(23): 4769.

25 Rice–Mycorrhizal Interaction Enhances the Biocontrol Efficiency through Integrated Approaches

Wiem Alloun
University Mentouri Constantine

Izdihar Ferhat
Ville Universitaire Ali Mendjeli

Hadjer Kecies
Abdelhafid Boussouf University Centre Mila

Aya Rehouma
Ville Universitaire Ali Mendjeli

Abdelkader Mahrouk
University of Mentouri Constantine

CONTENTS

DOI: 10.1201/9781003354086-25

25.1 INTRODUCTION

Chemical pesticides and fertilizers cause severe threats to global health and are the subject of progressive banishment. Thus, managing biotic and abiotic stresses in rice (*Oryza sativa* L), which lately has been intensified by climate changes and agricultural practices, has become one of the most concerning plant pathology and adaptability (Mitra et al. 2021a). Microbial inoculants represent promising sustainable agricultural technology for optimally preserving plant health, improving growth, and enhancing essential nutrient uptake (Razak and Gange 2021). The biocontrol capacity of a microbe can involve several mechanisms: antibiosis, fungal cell wall-degrading enzymes, induced systemic resistance, and iron chelation.

Thus, an active competition for the nutritional resources of the rhizosphere may be implicated in the biocontrol process. However, the success of biocontrol may depend on suitable formulations and the survival rate of the microbial agents. Arbuscular mycorrhizal fungi (AMF) have recently gained interest due to their specific interactions with their host plants. They colonize plant roots via intracellular fungal structures, namely, arbuscules (Rillig et al. 2002). AMF contribute to soil structure, stability, and function through increased growth and mass from a host's root system. Their role is in biotic stress management and abiotic stress alleviation, resulting in enhanced crop yields (Mitra et al. 2022; Dastogeer et al. 2020).

Some studies suggest that the colonization of plant roots is justified by the fungi's competition for nutrients mobilized by their plant root exudates and nutrients essential for their growth. By colonizing their host rhizosphere, they actively shape the microbial community in the surrounding soil, known as the rhizosphere effect. These microorganisms also induce the plant's systemic resistance by activating different patterns (Mitra et al. 2021b).

In this chapter, relevant knowledge on AMF-mediated stress alleviation was integrated cohesively to explain (i) the beneficial effects of host plant root colonization by AMF, (ii) their symbiosis of rice plants as a biosystem defining their high functionality under suboptimal conditions, and finally, (iii) the mechanism by which AMF enhance plant resistance to biotic stress, featuring them as promising bio-inoculants for growth phytostimulation and biocontrol effects.

25.2 RICE PRODUCTION AND BIOTIC STRESSORS

Rice constitutes the world's first most-produced agricultural crop in India and second in the world. Its production is geographically centered in Western and Eastern Asia. As a staple food, with 756 million tons produced globally in 2019, rice is a vital part of nutrition in much of Asia, Latin America, Africa, and the Caribbean. Figure 25.1 summarizes the most critical countries responsible for the bulk of global rice production. China and India conquered the top of the chart with 389 million tons of rice combined, accounting for more than half of the worldwide output. Most of the top producers are in Asia, except for Brazil (Wallach 2022).

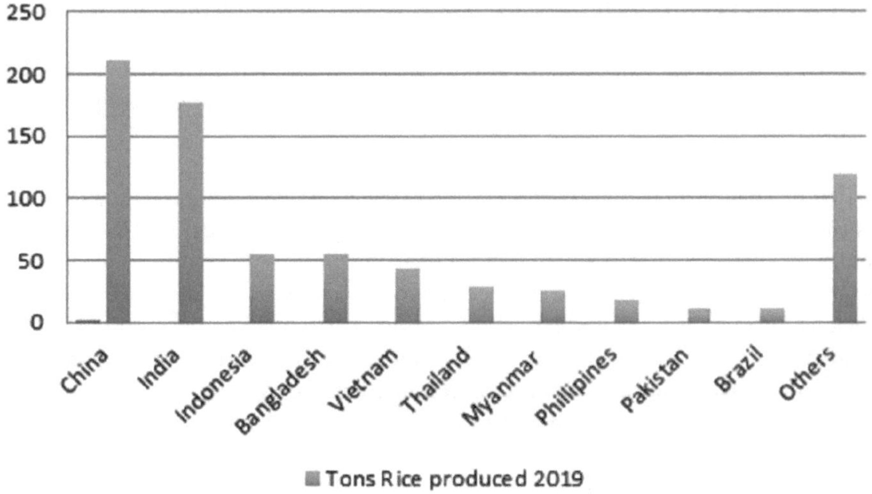

FIGURE 25.1 The great rice producers in the world. (Wallach, 2022.)

Nevertheless, substantial annual losses in rice yield are caused by various pathogens such as fungi, bacteria, and viruses that infect rice plants, causing severe diseases at all stages of rice development (Do 2022). *Magnaporthe grisea* causes the primary rice pandemic diseases (rice blast), *Rhizoctonia solani* (sheath blight), and the rice tungro virus (tungro disease).

25.3 AMF-INTEGRATED APPROACH FOR THE BIOCONTROL OF RICE PATHOGENS

Regardless of their demerits, agrochemicals remain the primary disease management practice for controlling rice soilborne pathogens. Breeding, genetic manipulation, selection of resistant cultivars, and inoculation of natural antagonists against these pathogens have recently emerged as alternatives to enhance plant protection. Among these potential biocontrol agents, AMF are the most globally explored biofertilizers (Mitra et al. 2020; Pirttilä et al. 2021).

25.3.1 Specific Symbiosis between AMF and Their Host Plants

AMF are biotrophic fungi, lacking essential metabolic enzymes, thus depending on their host root exudates rich in carbohydrates, and in turn, they provide fitness and control from parasites and harmful microorganisms (Sun et al. 2019). AMF establish close symbiosis with about 80%–90% of vascular plant species and 90% of crop plants (Smith and Read 2010). This symbiosis with their host is a mutually beneficial interaction between the fungi and the plants, promoting global phosphate cycling in terrestrial ecosystems and facilitating plant mineral acquisition. AMF are recognized as obligate symbionts requiring root colonization to complete a life cycle through the production of propagules and asexual spores. Root colonization starts with the penetration of plant root cortical cells and the formation of the arbuscules for nutrient exchange (Mitra et al. 2020).

The implication of phytohormones in the interactions between AMF and plants has been suggested recently (Liao et al. 2018). Nonetheless, there is only limited information about the direct effect of phytohormones on AMF growth and reproduction. Additionally, not all phytohormones positively impact interactions between AMF and their plant hosts. For instance, Mitra et al. (2021a) reviewed the role of strigolactones (SL) as signal molecules in regulating rice plant root growth and assisting fungal symbiotic relationships. SL mediate the symbiotic relationship between plants and their fungi symbionts, which can eventually increase the P acquisition from the surrounding soil (Tanaka et al. 2022).

25.3.2 AMF Responses to Rice Pathogens and Biocontrol Effect

AMF effectively control fungal and bacterial pathogens such as *Rhizoctonia solani*, *Pythium ultimum*, and *Phytophthora* (Mitra et al. 2021a). AMF and rice host plants' interaction protects them from biotic stresses by increasing natural antagonism against soilborne phytopathogens and reducing disease severity (Mitra et al. 2021b). For instance, Nasir et al. (2019) reported an acquired resistance of rice plants against *M. oryzae* through the induction of metabolic pathways and cellular processes. Furthermore, rice brown spot disease caused by *Cochliobolus miyabeanus* was effectively reduced by combining AMF with the two bioagents *Pseudomonas fluorescens* and *Trichoderma viride in vivo* and in field conditions (Saleh and El-Akshar 2020). Pandit et al. (2022) reported the broad-spectrum effective control of the biocontrol properties of AMF against fungal root pathogens, such as *Colletotrichum*, *Alternaria*, *Erysiphe*, *Gaeumannomyces*, *Macrophomina*, *Botrytis*, *Rhizoctonia*, *Fusarium*, *Cylindrocladium*, *Sclerotium*, and *Verticillium*.

25.4 PERFORMANCE OF AMF IN STRESS ALLEVIATION AND NUTRIENT DEFICIENCY

The symbiotic relationships between AMF and the host plant affect the performance of the plant under suboptimal conditions and, subsequently, the production of metabolites, including sugars, proteins, phenols, tannins, and flavonoids, expressed as crop yield (Posta and Duc 2020) by increasing macronutrients' availability and uptake (Dastogeer et al. 2020). Under individual biotic and abiotic stresses or combined multi-stresses, including drought, salinity, heavy metals and toxic substances, nutrient deficiency, or microbial pathogens, AM-associated plants exhibited better growth than non-inoculated plants (Mitra et al. 2022) (Table 25.1).

TABLE 25.1
AMF Contribution in Rice Biotic and Abiotic Stress Alleviation

Stress	AMF Strains	Responses of Plants Post-AMF Inoculation	References
Metal stress arsenate (AsV) or arsenite (AsIII)	*Rhizophagus irregularis*	Improve biomass and relative chlorophyll content. Sustain higher rice photosynthesis when As is present as AsV	de Andrade et al. (2015)
Selenium uptake and accumulation	*Glomus versiforme*, *Funneliformis mosseae*	Enhanced organic selenium accumulation in rice grain	Chen et al. (2020)
Cd toxicity	*F. mosseae*	Decreased cadmium uptake	Chen et al. (2019)
	Solanum nigrum + *F. mosseae*	Cd accumulation in various rice plants parts	Yang et al. (2021)
Saline stress	Combination of *Acaulospora laevis* and *Gigaspora margarita*	Increased rice yield	Parvin et al. (2020)
	Combination of *R. irregularis* and bacteria (*P. fluorescens* + *P. putida*)	Increased grain yield and decreased salinity	Norouzinia et al. (2020)
Spodoptera frugiperda	*R. irregularis, F. mosseae,*	Increased chlorophyll content, increased shoot K+/Na+ ratio and a low shoot Na+/root Na+ ratio along with improved root biomass, spikelet fertility, and grain production compared to non-inoculated control plants	Bernaola et al. (2018)
Rhizoctonia solani	*G. deserticola, R. fasciculatum, Sclerocystis dussii,* and *G. microaggregatum*		
High CO_2 concentration	Combination of native Glomerales species	Improved grain yield by 25.08% and increased phosphorus uptake by 18.2% and nitrogen by 49.5% at 700 ppm CO_2 concentration	Panneerselvam et al. (2019)

25.5 CONCLUSION

AMF have been historically monumental in the development of plant evolution and are critical symbiotic organisms for agriculture with crop growth promotion and increase of soil stability and quality. Plants colonized by AMF can effectively deal with various environmental stress factors that induce plant growth and rice production. Shreds of evidence on the stress alleviation potential of AMF are growing. Undoubtedly, AMF constitute an alternative to synthetic fertilizers and other chemicals, thus promoting organic farming. AMF-driven growth improves rice crop productivity, which helps meet the consumption needs of the world's rising population. Further research should focus on identifying genes that regulate the growth and development mediated by AMF under stressful conditions and the specific signaling at the basis of the AMF-plant symbiotic association and the complex cellular pathways. From this perspective, the exploration of AMF as plant growth biostimulant and biocontrol agents opens the doors to a new generation of organic fertilizers and pesticides for sustainable organic rice production.

REFERENCES

Bernaola, Lina, Marco Cosme, Raymond W. Schneider, and Michael Stout. 2018. "Belowground Inoculation with Arbuscular Mycorrhizal Fungi Increases Local and Systemic Susceptibility of Rice Plants to Different Pest Organisms." *Frontiers in Plant Science* 9: 1–16. https://doi.org/10.3389/fpls.2018.00747.

Chen, Xue, Zengyu Zhang, Minghua Gu, Hong Li, Md. Jahidul Islam Shohag, Fangke Shen, Xueli Wang, and Yanyan Wei. 2020. "Combined Use of Arbuscular Mycorrhizal Fungus and Selenium Fertilizer Shapes Microbial Community Structure and Enhances Organic Selenium Accumulation in Rice Grain." *Science of the Total Environment* 748. https://doi.org/10.1016/j.scitotenv.2020.141166.

Chen, Xun Wen, Li Wu, Na Luo, Ce Hui Mo, Ming Hung Wong, and Hui Li. 2019. "Arbuscular Mycorrhizal Fungi and the Associated Bacterial Community Influence the Uptake of Cadmium in Rice." *Geoderma* 337: 749–57.

Dastogeer, Khondoker M. G., Mst Ishrat Zahan, Md Tahjib-Ul-Arif, Mst Arjina Akter, and Shin Okazaki. 2020. "Plant Salinity Tolerance Conferred by Arbuscular Mycorrhizal Fungi and Associated Mechanisms: A Meta-Analysis." *Frontiers in Plant Science*. Frontiers Media S.A. https://doi.org/10.3389/fpls.2020.588550.

de Andrade, Sara Adrian Lopez, Adilson Pereira Domingues, and Paulo Mazzafera. 2015. "Photosynthesis Is Induced in Rice Plants that Associate with Arbuscular Mycorrhizal Fungi and Are Grown under Arsenate and Arsenite Stress." Chemosphere 134: 141–49. https://doi.org/10.1016/j.chemosphere.2015.04.023.

Do, Quang Trung. 2022. "Antagonistic Activities of Endophytic Bacteria Isolated from Rice Roots against the Fungus Magnaporthe Oryzae, a Causal of Rice Blast Disease." *Egyptian Journal of Biological Pest Control* 32(1): 1–11. https://doi.org/10.1186/s41938-022-00571-1.

Liao, Dehua, Shuangshuang Wang, Miaomiao Cui, Jinhui Liu, Aiqun Chen, and Guohua Xu. 2018. "Phytohormones Regulate the Development of Arbuscular Mycorrhizal Symbiosis." *International Journal of Molecular Sciences*. Multidisciplinary Digital Publishing Institute (MDPI). https://doi.org/10.3390/ijms19103146.

Mitra, Debasis, Beatriz Elena Guerra Sierra, Bahman Khoshru, Sergio De Los Santos Villalobos, Claudia Belz, Priya Chaudhary, Faride Noroozi Shahri, et al. 2021a. "Impacts of Arbuscular Mycorrhizal Fungi on Rice Growth, Development, and Stress Management with a Particular Emphasis on Strigolactone Effects on Root Development." *Communications in Soil Science and Plant Analysis* 52(14): 1591–1621. https://doi.org/10.1080/00103624.2021.1892728.

Mitra, Debasis, Boya Saritha, Edappayil Janeeshma, Poonam Gusain, Bahman Khoshru, Fatma A. Abo Nouh, Anju Rani, et al. 2022. "Arbuscular Mycorrhizal Fungal Association Boosted the Arsenic Resistance in Crops with Special Responsiveness to Rice Plant." *Environmental and Experimental Botany*. https://doi.org/10.1016/j.envexpbot.2021.104681.

Mitra, Debasis, Navendra Uniyal, Periyasmay Panneerselvam, Ansuman Senapati, Arakalagud Nanjundaiah Ganeshamurthy, Vikas Kumar, and Divya Jain. 2020. "Role of Mycorrhiza and Its Associated Bacteria on Plant Growth Promotion and

Nutrient Management in Sustainable Agriculture." International Journal *of* Life Sciences and Applied Sciences 1(1): 1–1. https://www.ijlsas.com/?mno=302644881.

Mitra, Debasis, Rihab Djebaili, Marika Pellegrini, Bhaswatimayee Mahakur, Aniruddha Sarker, Priya Chaudhary, Bahman Khoshru, et al. 2021b. "Arbuscular Mycorrhizal Symbiosis: Plant Growth Improvement and Induction of Resistance under Stressful Conditions." *Journal of Plant Nutrition*. Taylor & Francis. https://doi.org/10.1080/01904167.2021.1881552.

Nasir, Fahad, Shaohua Shi, Lei Tian, Chunling Chang, Lina Ma, Xiujun Li, Yingzhi Gao, and Chunjie Tian. 2019. "Strigolactones Shape the Rhizomicrobiome in Rice (*Oryza Sativa*)." Plant Science 286: 118–33. https://doi.org/10.1016/j.plantsci.2019.05.016.

Norouzinia, Firouz, Mohammad Hossein Ansari, Hashem Aminpanah, and Saeed Firozi. 2020. "Alleviation of Soil Salinity on Physiological and Agronomic Traits of Rice Cultivars Using Arbuscular Mycorrhizal Fungi and Pseudomonas Strains under Field Conditions." *Revista De Agricultura Neotropical* 7 (1): 25–42. https://doi.org/10.32404/rean.v7i1.4042.

Pandit, Manisha Arora, Jitendra Kumar, Saloni Gulati, Neeru Bhandari, Poonam Mehta, Roma Katyal, Charu Dogra Rawat, Vachaspati Mishra, and Jasleen Kaur. 2022. "Major Biological Control Strategies for Plant Pathogens." *Pathogens*. Multidisciplinary Digital Publishing Institute. https://doi.org/10.3390/pathogens11020273.

Panneerselvam, Periyasamy, Sowarnalisha Sahoo, Ansuman Senapati, Upendra Kumar, Debasis Mitra, Chidambaranathan Parameswaran, Annamalai Anandan, Anjani Kumar, Afrin Jahan, and Amaresh Kumar Nayak. 2019. "Understanding Interaction Effect of Arbuscular Mycorrhizal Fungi in Rice under Elevated Carbon Dioxide Conditions." *Journal of Basic Microbiology* 59(12): 1217–28. https://doi.org/10.1002/jobm.201900294.

Parvin, Shanaz, Maarten Van Geel, Tanzima Yeasmin, Erik Verbruggen, and Olivier Honnay. 2020. "Effects of Single and Multiple Species Inocula of Arbuscular Mycorrhizal Fungi on the Salinity Tolerance of a Bangladeshi Rice (*Oryza Sativa* L.) Cultivar." *Mycorrhiza* 30. https://doi.org/10.1007/s00572-020-00957-9.

Pirttilä, Anna Maria, Habibollah Mohammad Parast Tabas, Namrata Baruah, and Janne J. Koskimäki. 2021. "Biofertilizers and Biocontrol Agents for Agriculture: How to Identify and Develop New Potent Microbial Strains and Traits." *Microorganisms*. MDPI. https://doi.org/10.3390/microorganisms9040817.

Posta, Katalin, and Nguyen Hong Duc. 2020. "Benefits of Arbuscular Mycorrhizal Fungi Application to Crop Production under Water Scarcity." *Drought Detection and Solution*, 1, 1–26.

Razak, Nadia Ab, and Alan C Gange. 2021. "Multitrophic Interactions between Arbuscular Mycorrhizal Fungi, Foliar Endophytic Fungi and Aphids." *Microbial Ecology*. https://doi.org/10.1007/s00248-021-01937-y.

Rillig, Matthias C., Sara F. Wright, and Valerie T. Eviner. 2002. "The Role of Arbuscular Mycorrhizal Fungi and Glomalin in Soil Aggregation: Comparing Effects of Five Plant Species." *Plant and Soil* 238(2): 325–33. https://doi.org/10.1023/A:1014483303813.

Saleh, Mona, and Yasser El-Akshar. 2020. "Integration between Arbuscular Mycorrhizal Fungi, Bacterial and Fungal Bioagents for Controlling Rice Brown Spot Disease." *Egyptian Journal of Phytopathology* 48(1): 81–93. https://doi.org/10.21608/ejp.2020.47059.1011.

Smith, Sally E., and David J. Read. 2010. *Mycorrhizal Symbiosis*. Academic Press: Cambridge, UK.

Sun, Xuepeng, Wenbo Chen, Sergey Ivanov, Allyson M. MacLean, Haley Wight, Thiruvarangan Ramaraj, Joann Mudge, Maria J. Harrison, and Zhangjun Fei.

2019. "Genome and Evolution of the Arbuscular Mycorrhizal Fungus Diversispora Epigaea (Formerly Glomus Versiforme) and Its Bacterial Endosymbionts." *New Phytologist* 221(3): 1556–73. https://doi.org/10.1111/nph.15472.

Tanaka, Sachiko, Kayo Hashimoto, Yuuki Kobayashi, Koji Yano, Taro Maeda, Hiromu Kameoka, Tatsuhiro Ezawa, Katsuharu Saito, Kohki Akiyama, and Masayoshi Kawaguchi. 2022. "Asymbiotic Mass Production of the Arbuscular Mycorrhizal Fungus Rhizophagus Clarus." *Communications Biology* 5(1). https://doi.org/10.1038/s42003-021-02967-5.

Wallach, Omri, and Amy Realey. 2022. "Visualizing the World's Biggest Rice Producers." *Visual Capitalist*. https://www.visualcapitalist.com/worlds-biggest-rice-producers/.

Yang, Xu, Junhao Qin, Jiachun Li, Zhenai Lai, and Huashou Li. 2021. "Upland Rice Intercropping with Solanum Nigrum Inoculated with Arbuscular Mycorrhizal Fungi Reduces Grain Cd While Promoting Phytoremediation of Cd-Contaminated Soil." *Journal of Hazardous Materials* 406: 124325. https://doi.org/10.1016/j.jhazmat.2020.124325.

Index

Note: **Bold** page numbers refer to tables and *italic* page numbers refer to figures.